# AUTOMATION 4.0

## Object-Oriented Development of Modular Machines for Digital Production

# AUTOMATION 4.0

## Object-Oriented Development of Modular Machines for Digital Production

**Thomas Schmertosch**

Leipzig University of Applied Sciences, Germany

**Markus Krabbes**

Merseburg University of Applied Sciences, Germany

**Christian Zinke-Wehlmann**

Institute for Applied Informatics at the Leipzig University, Germany

**W World Scientific**

NEW JERSEY · LONDON · SINGAPORE · BEIJING · SHANGHAI · TAIPEI · CHENNAI

*Published by*

World Scientific Publishing Co. Pte. Ltd.

5 Toh Tuck Link, Singapore 596224

*USA office:* 27 Warren Street, Suite 401-402, Hackensack, NJ 07601

*UK office:* 57 Shelton Street, Covent Garden, London WC2H 9HE

**Library of Congress Cataloging-in-Publication Data**
Names: Schmertosch, Thomas, author. | Krabbes, Markus, author. |
    Zinke-Wehlmann, Christian, author.
Title: Automation 4.0 : object-oriented development of modular machines for digital production /
    Thomas Schmertosch, Leipzig University of Applied Sciences, Germany,
    Markus Krabbes, Merseburg University of Applied Sciences, Germany,
    Christian Zinke-Wehlmann, Institute for Applied Informatics at the Leipzig University, Germany.
Description: Toh Tuck Link, Singapore ; Hackensack, NJ : World Scientific Publishing Co. Pte. Ltd,
    [2025] | Includes bibliographical references and index.
Identifiers: LCCN 2024029683 | ISBN 9789811297014 (hardcover) |
    ISBN 9789811297021 (ebook for institutions) | ISBN 9789811297038 (ebook for individuals)
Subjects: LCSH: Manufacturing processes--Automation. | Object-oriented methods
    (Computer science)--Industrial applications. | Industry 4.0.
Classification: LCC TS183 .S355 2025 | DDC 670.42/7--dc23/eng/20250109
LC record available at https://lccn.loc.gov/2024029683

**British Library Cataloguing-in-Publication Data**
A catalogue record for this book is available from the British Library.

For any available supplementary material, please visit
https://www.worldscientific.com/worldscibooks/10.1142/13955#t=suppl

Desk Editors: Kannan Krishnan/Amanda Yun

Typeset by Stallion Press
Email: enquiries@stallionpress.com

# Preface

Since the German government coined the term Industry 4.0, there has hardly been an engineering conference without at least one presentation on this topic. The entire series of topics, online offerings, and specialist books have been devoted to Industry 4.0 from different perspectives. And indeed, we are on the threshold of a new form of industrialization. But we have already crossed it and have been doing so for a long time.

The Gabler Business Dictionary defines Industry 4.0 as follows:

*"Industry 4.0" is a marketing term that is also used in science communication and stands for a "future project" of the German government. The so-called fourth industrial revolution is characterized by the individualization and hybridization of products and the integration of customers and business partners into business processes.*

This definition contains the core topics of "individualization of products" and the "change in business processes". More information on this process can be found on the official website of the German government, the *Industry 4.0 platform:*

*In Industry 4.0, production is interlinked with state-of-the-art information and communication technology. The driving force behind this development is the rapidly increasing digitalization of the economy and society.*

Digitalization is thus declared to be the driving force. And indeed, the technical world has changed profoundly since the market launch of smartphones and the like. The self-image of customers who prefer to shop online rather than in the corner store next door, growing demands on availability, logistics, quality, and, finally, extreme price sensitivity cannot leave the production environment unscathed. The authors therefore focus on the manufacture and operation of production equipment and which requirements are put upon them by such megatrends. To get closer to the answer, it is worth reading further in the publications of the *Industry 4.0 platform* (as this publication also provides comprehensive answers to many other questions relating to the topic):

*The technical basis for this are intelligent, digitally networked systems that enable largely self-organized production: People, machines, systems, logistics and products communicate and cooperate directly with each other in Industry 4.0. Production and logistics processes between companies in the same production process are intelligently interlinked in order to make production even more efficient and flexible.*

This poses a central challenge:

People, machines, systems, logistics, and products should be able to communicate with each other. But haven't they already been doing this for some time?

Machines have always had communication options adapted to their requirements in terms of scope and equipment. In the beginning, these were mechanical control elements such as levers, adjusting screws, or handwheels. At first, operators could only detect the machine's status by observing and measuring as well as with a great deal of experience. With increasing electrification, switches, and illuminated displays were added and, with the introduction of electronic systems, increasingly qualitative instruments such as potentiometers and illuminated scales.

After all, since the introduction of PC-based hardware and software in the operating technology of machines and systems, we have seen an ever deeper integration of digitized methods and mechanisms, as we know them from smartphones and tablets. What is now called the *Human–Machine Interface* (HMI) often looks confusingly similar to these devices and offers more and more convenience. One example of this is multi-touch technology, which has become established for smartphones and has also revolutionized the operation of machines. However, as the possibilities increase, so do the demands on the design of visual representations — a topic that has introduced the new profession of a usability engineer. Additionally, this book also deals with the direct communication of machines and production systems in the digital production environment.

The networking of production and control levels has been state of the art since the early 1990s. After all, the data from production equipment is used for effective production control and therefore makes a significant contribution to reducing unit costs. Even though new communication systems from the office world such as Ethernet and WLAN have long since found their way onto the shop floor, inter-machine communication still follows the same principles as it did in the early days.

This is because many manufacturers of automation technology still have their own ideas about data communication. Therefore, data collectors, bridges, and so-called interpreters populate the factory halls, whose sole task is to process the data in such a way that other machines are able to understand it right up to the production control system. That makes it a lucrative business for engineering firms and software service providers that will hopefully soon be a thing of the past.

Unfortunately, industry-specific communication standards such as *PackML* in the packaging industry, *JDF* in the printing industry, or *Euromap* in the plastics industry have only provided moderate relief. However, Industry 4.0 demands that ALL components involved in the manufacturing process communicate DIRECTLY with each other. This requirement in particular is crucial for the production of individualized products, so we will deal with this topic in detail.

The Industry 4.0 platform describes another feature as follows: *(...) intelligent value chains are created that also include all phases of the product life cycle — from the idea of a product to development, production, use, maintenance, and recycling. In this way, customer requirements can be taken into account from the product idea through to recycling, including the associated services. This makes it easier for companies to produce customized products according to individual customer requirements. The individual manufacture and maintenance of products could become the new standard.*

This means nothing other than *batch size 1* instead of series and mass production. But *batch size 1* can only be successful if the individualized product is of the same quality and does not require higher costs than the comparable series product. This is why the Industry 4.0 platform continues:

*On the other hand, production costs can be reduced despite individualized production. By networking the companies in the value chain, it is no longer possible to optimize just one production step, but the entire value chain. (...) Production processes can be controlled across companies in such a way that they save resources and energy.*

Since the authors believe that, despite individualization, production costs not only CAN but MUST fall, they focus intensively on the question of how this can be achieved using state-of-the-art automation technology. Not only the individualized end product is considered but also the individual production equipment, as both influence the design of an automation system in different ways. Consequently, this book focuses on the resulting aspects and solution approaches. In this way, *Automation 4.0* aims to show both experienced and future engineers the requirements and solutions for automation concepts that will make a production system fit for the future and Industry 4.0.

While dealing with this comprehensive topic, we repeatedly became aware of the extraordinary intensity with which numerous institutes, associations, and manufacturers are tackling these challenges. Time and again, we found that new approaches, solutions, and standards were emerging that had not even been mentioned at the beginning of our research. A typical example of this is the *OPC UA* bus protocol in combination with *Time Sensitive Networking*. All statements made in this book refer to the technical status of March 2018 and have been updated in the 2nd edition to August 2023. The same applies to the statements on *Industry 4.0-compliant communication*, the standardization process of which will continue for some time. We were therefore keen to include statements and opinions from experts involved in various committees in addition to the numerous literature references to provide the reader with the latest research.

Special thanks go to Dr. Heiko Koziolek from the ABB AG research center and member of the VDI/VDE-GMA expert committee "Industrie 4.0". Especially because of him, we were able to compare the numerous publications of the Industry 4.0 platform with the latest developments and consolidate them in this book.

The same applies to Mr. Sebastian Sachse, who represented B&R Industrial Automation GmbH on several international committees as an employee of the Open Automation Technologies Business Unit until 2020. Thanks to his advice, it was also possible to add the latest facts and figures to the presentations on the topic of communication.

We would like to thank Dipl.-Ing., Dipl.-Wirtsch.-Ing. Peter von Dreusche, Head of Electrotechnical Engineering Unit at Trützschler GmbH & Co. KG, for the intensive exchange of ideas on the subject of *object-oriented modularization*. It is thanks to him that we are able to offer this method in such a well-founded and practical way. We would also like to thank Prof. Dr. rer. nat. Matthias Krause for the exchange of ideas and expert advice on this topic.

We would also like to thank the numerous employees of B&R Industrial Automation GmbH. It would not have been possible to comprehensively present the practical topics without their support and expertise. In particular, we would like to thank Mr. Franz Kaufleitner, Product Manager for Integrated Safety Technology at B&R, for his numerous tips and support on the subject of *safety technology* and Mr. Franz Enhuber, Head of the B&R Automation Academy. The several discussions with him were particularly helpful in analyzing market requirements and development trends in modern automation technology. Finally, we would like to thank the entire B&R marketing team and its corporate editor, Mr. Stefan Hensel, for providing the numerous images.

We would also like to thank the german companies Fomanu AG Neustadt, Sick AG Waldkirch, Pilz GmbH & Co. KG Ostfildern, Buchbinderei Johst in Wermsdorf and many others for providing photographic material.

We would like to thank Ms. Dipl.-Ing. (FH) Franziska Kaufmann, Mr. Manuel Leppert, and all employees of Fachbuchverlag Leipzig for excellent cooperation on the 1st edition. Finally, we would like to thank the World Scientific Group Company for editing this edition and for the excellent cooperation with the editors, Amanda Yun and Kannan Krishnan.

Special thanks go to my co-author Prof. Dr.-Ing. Markus Krabbes, the long-standing vice-president for Research at Leipzig University of Applied Sciences (HTWK) and current president of Merseburg University of Applied Sciences. He gave me the initial impetus for this book and contributed a significant aspect by writing Chapter 3, *Project Planning of Modular Machines*. I would also like to thank my co-author Dr.-Ing. Christian Zinke-Wehlmann from the Institute for Applied Computer Science (InfAI) at the University of Leipzig, where he is head of the competence center for KMI and conducts research in the field of applied artificial intelligence.

What is an author without the active support of his family? Nothing. With this in mind, I would like to thank my family in particular for their understanding and patience in supporting and enabling me to work on this book.

Thomas Schmertosch
*Leipzig (Germany)*
2025

# About the Authors

**Thomas Schmertosch** has been an honorary professor at Leipzig University of Applied Sciences (HTWK Leipzig), Germany, since 2014 and is responsible for the subject areas "Components of Automation" and "Modular Automation Systems". Born in Leipzig in 1952, he studied and graduated as a cyberneticist at the Otto-von-Guericke University Magdeburg (OVGU), Germany, and worked as such in crane and printing machine construction and for a leading manufacturer of automation technology until his retirement. Since 2016, he has been working as a freelance author and consulting engineer in the field of automation technology and Industry 4.0.

**Markus Krabbes** has held a professorship in the field of Information Systems at Leipzig University of Applied Sciences (HTWK Leipzig), Germany, since 2003. After two terms as Vice Rector for Research at HTWK Leipzig, Markus Krabbes has been Rector of Merseburg University of Applied Sciences since 2022. Born in Leipzig in 1970, he studied at the former Leipzig University of Technology (TH Leipzig), Germany, and obtained his doctorate from the Faculty of Electrical Engineering and Information Technology at the Otto-von-Guericke University Magdeburg (OVGU), Germany, and worked as a scientist at the Fraunhofer Institute for Machine Tools and Forming Technology in Chemnitz, Germany. In its Machine and Process Informatics department, he worked on mechatronic tasks on machine tools.

**Christian Zinke-Wehlmann** is Head of the Competence Centre for Artificial and Human Intelligence (KMI) at the Institute for Applied Computer Science (InfAI), at Leipzig University, Germany, where he operates at the interface of digital technologies, education, work, and services. Born in Köthen in 1983, he studied sociology at Leipzig University, where he completed his doctorate under the Chair of Business Information Systems in Computer Science. With his multidisciplinary expertise, he has been dedicated to researching and solving complex socio-technical challenges in these areas since 2012.

# Acknowledgments

This work is partly funded under the funding measure "Future of Work: Regional Competence Centers for Work Research — Artificial Intelligence" in the program "Innovations for Tomorrow's Production, Services and Work" of the Federal Ministry of Education and Research (BMBF), Bonn, Germany, and supervised by the Project Management Agency Karlsruhe (PTKA) Fund. No.: 02L19C500.

# Contents

# List of Figures

# List of Tables

# Chapter 1

# Automation 4.0: Requirements and Prospects

Before we shift our focus to the requirements for automation technology, this chapter analyzes the requirements stemming from the definition of Industry 4.0. Additionally, there are general trends the mechanical and plant engineering industry must fundamentally address, from which important objectives for the design of automation systems can be derived.

## 1.1 Perception of Industry 4.0

As part of the fourth industrial revolution, a virtual marketplace is emerging where the production process is negotiated between machines offering their services and components capable of controlling their own machining process. This vision crosses old boundaries: Companies are forging new value chains by integrating business partners, suppliers, and customers into the production process. This integration allows customers to monitor their order online at any time and submit processing requests; it enables suppliers to ascertain real-time inventory and permits shipping companies to confirm the pickup dates while production is ongoing [1].

Moreover, as information technology, facilitated by the Internet, knows no borders, Industry 4.0 has become a global phenomenon. Megatrends such as increasing price and quality consciousness, global availability, and unlimited communication are reshaping markets and fostering alternative lifestyles. The consumer paradigm is shifting from ownership to access, with products being rented and/or highly personalized.

Advanced production and logistics processes enable the manufacture of standard cars in millions of distinct configurations on a single assembly line. This is complemented by services that integrate with other mobility offerings or leasing models as integral product components.

*Looking ahead, customers will not merely choose a product but a business model* [2]. In other words, *the fourth industrial (r)evolution particularly emphasizes the individualization and hybridization of products (through the coupling of production and services)* [3], challenging the norms of traditional mass production — a Herculean task.

There are numerous examples of this, as batch size one has already made its way into the real world. The photo book you design yourself on your PC at home (Figure 1.1) is as unique as the recipe for a personalized muesli mix. Similarly, pre-packaged medication sets for a week for senior citizens not only provide optimal therapy but also reduce packaging

**Figure 1.1.**   The photo book: Typical representative of individualized products in batch size one.

material and transport volume, a beneficial side effect. The uniqueness of these products is their unique selling point, enhancing their marketability. This shift is supported and driven by innovative technologies that not only offer new functionalities but also make individualization an almost incidental benefit.

In-mold labeling used in the production of plastic items, such as yogurt pots, bottles, and much more, is one such technology. This process involves placing a printed polypropylene label inside a mold, into which plasticized plastic is injected. The plastic merges with the label and assumes the mold's shape as it cools, resulting in a unified product with enhanced properties and functionalities (Figure 1.2). For instance, "*(...) in-mold labels are resistant to moisture and temperature and are therefore also suitable for frozen products and the degree of freshness can even be measured. In addition, there are improved mechanical properties such as scratch, shrink and tear resistance and much more*" [4, translated].

The upstream offset printing of labels is particularly noteworthy. Not only does it facilitate the printing of high-resolution color images,

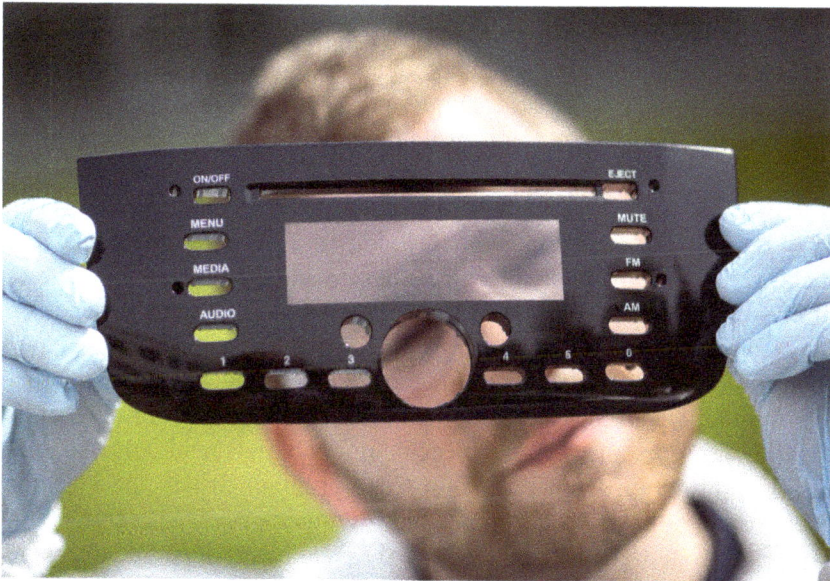

**Figure 1.2.**  Individual labeling and embedded sensors give plastic front panels new functional properties.

*Source*: Plastics Technology Institute at Ilmenau University of Technology.

transforming the packaging with a completely new aesthetic, but it also enables precise individualization of products and their packaging. For example, a special muesli mix in a box with the inscription "Mummy's special 50th birthday edition from Lisa and Tim", supplemented with a photo of the gift-givers, will undoubtedly be a standout gift. Similarly, a medication set personalized with the recipient's name in large letters and detailed intake instructions adds significant value. If such customizations can be achieved without significant additional costs, real added value and business success are guaranteed.

However, the challenge lies here. For generations, it has been understood that unit costs decrease with mass and series production, a concept that led to the begin of assembly line production and the division of labor in the early 20th century. Today, we refer to this development as the 2nd industrial revolution. It is followed by the 3rd industrial revolution, with the introduction of electronic (programmable logic controller) automation, primarily aiming to expedite and improve the efficiency of repetitive tasks over human capabilities. The resulting job displacement is a significant concern, manageable only through societal interventions and adjustments (e.g. re-skilling).

Until the end of the 20th century, machines were predominantly designed to manufacture a single product type, such as entire production lines for a specific relay or a unique light source. When these products were phased out after several million units, the machines were often discarded.

In contrast, while series machines like packaging machines can adapt to different product parameters, changing to a new format or material requires forethought in the machine's design and entails varying degrees of conversion effort. The diversity of the resulting products depends on the machine's setup and parameterization capabilities.

Conversely, a modern machine tool is inherently suited for producing individualized pieces. Designed and equipped accordingly, it is capable of processing a single piece completely independently. All necessary information is encoded in a CNC program, which the machine operator edits directly on the machine's control panel. Alternatively, it can be generated via the designer's CAD system with a simple click and transferred to the machine through a network or storage medium. Thus, the conversion process involves merely importing the machining program and loading the tool magazine, often automated with robots and other feeding systems.

The same applies to the deposition technology used in 3D printers. A plastic injection molding machine (Figure 1.3) can produce several products, such as gear wheels, plastic boxes, or entire dustbins, in a single operation. A complete operation, involving liquefying (plasticizing) the material, injecting it into a mold (the injection mold), cooling it, and removing it (with a removal robot), usually takes only seconds. This leaves the 3D printer no chance in terms of time. So why does it exist at all?

Here too, the key to understanding lies in the production method. An injection mold represents a significant investment, requiring dimensional stability under extreme pressures (1,000 bar is not uncommon) and high temperatures, with precision down to the micrometer. The mold surfaces are highly polished, and various internals (cavities and cores) allow multicolors and even movable joints to be formed in a single operation. All this comes at a price, which quickly exceeds million dollars. This investment becomes cost-effective only when amortized over large production volumes, achieved through meticulously planned and continuously optimized production cycles, measured in tenths of a second.

This is the opportunity for 3D printers. Despite being significantly slower than plastic injection molding machines, they can produce individual workpieces of any shape and contour by simply uploading a digital file. Moreover, this flexibility extends to a wide range of materials,

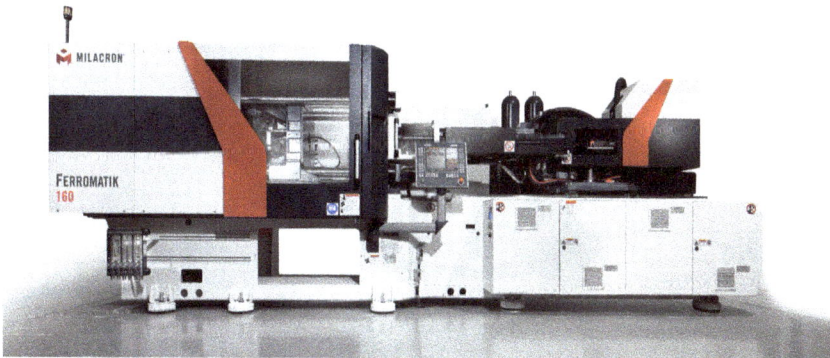

**Figure 1.3.** In a plastic injection molding machine, multi-colored products can be manufactured in a single operation.

*Source*: Ferromatik Milacron GmbH.

including plastics, metals, minerals, and even organics. Thanks to the elimination of tooling and the significantly simplified (and therefore cheaper) machine design, the unit costs are also within a very acceptable range even with the longer cycle times. Thus, 3D printers have become indispensable for prototype production, though they may not be suitable for mass production in the near future. Additionally, the technology of layer-by-layer deposition (additive manufacturing) enables the production of highly complex structures and shapes unachievable through conventional machining. This is even possible for highly stressed workpieces, such as blades for gas turbines [5]. These components are welded using laser metal deposition (Figure 1.4) and can withstand temperatures of over 1,250 °C at 13,000 revolutions per minute, thanks to a novel blade design featuring improved internal cooling geometries. As a result, 3D printers are already occupying the first niches in series production.

The examples clearly demonstrate that individualized products do not necessarily incur higher costs or longer production times than mass-produced items — provided that the accounting encompasses the entire

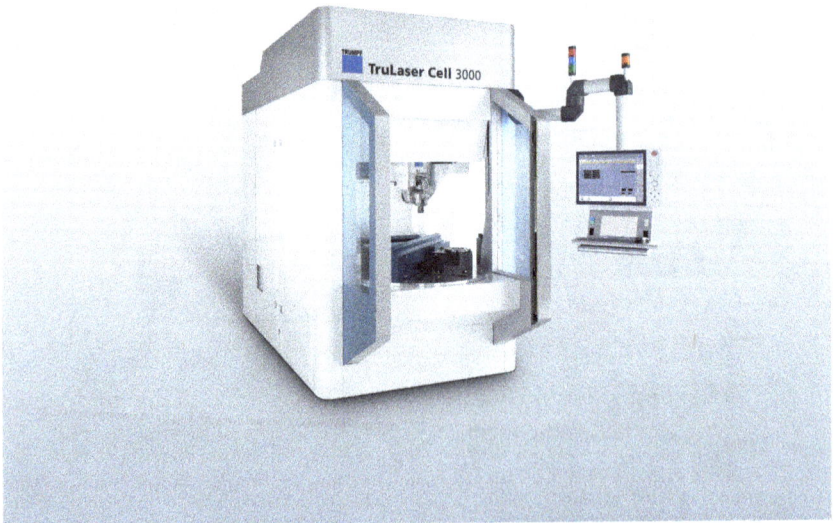

**Figure 1.4.**   3D laser machine for laser metal deposition.

*Source*: Trumpf GmbH & Co. KG.

production process. However, they also underline that batch size one *PLUS* lowest unit costs *PLUS* shortest unit times cannot be achieved at the same time (yet).

However, as engineers, accepting this limitation would fall short of our ambitions, and believing in a "one-size-fits-all" solution would be even less acceptable. That is also not the spirit of Industry 4.0. Our objective (and the aim of this book) is to explore a pathway that harmonizes the best of both worlds, integrating the benefits of various technologies and enhancing them with new innovations. By doing so, we strive to ensure that individualized series products are neither more time-consuming to produce nor more costly than standard mass-produced goods, while also transforming highly complex specialized machinery into versatile series production tools.

Beyond the scope of Industry 4.0, emerging trends are shaping machine concepts and their required automation in significant ways. Before diving into strategies and solutions of Automation 4.0, let's take a closer look at the most important of these trends.

## 1.2 Trends and requirements in mechanical and plant engineering

The work of developers and designers has always been determined by customer requirements. Trends can be derived from the sum of these requirements, which gradually establish themselves as ever-evolving standards. In our digital age, many of these requirements can only be met with automation and IT technologies. The most important trends are presented as follows.

### 1.2.1 End products determine the direction

The requirements for all types of goods have always been subject to the megatrends of human needs. Since the beginning of mechanization, this has naturally also applied to machines of all kinds. For instance, the desire for patterned fabrics led to the invention of new weaving machines, while the development of hygienic standards spurred the creation of vacuum packaging machines. The awareness that our raw material resources are limited led to the development of carbon fibers to save energy through lightweight construction and solar panels to generate energy in an environmentally friendly way. Further, machines and systems, along with household appliances and light bulbs, must reduce their energy consumption and ultimately be labeled accordingly.

It is evident that the end products' requirements — be it textiles, contact lenses, photo books, ready-made pizzas, or refrigerators — determine and drive the necessary production technologies. When designing and developing processing machines or production systems, the end product is the primary focus. This leads to the selection of raw materials, auxiliary materials, processes, and technologies, which must be reflected in the design and ultimately in the automation and service of the processing machines.

Yet, if development were confined solely to these specifications, the field of engineering would lack excitement. What renders the profession stimulating, and admittedly sometimes challenging, are the multiple constraints, laws, and regulations that often significantly increase the design effort.

Just imagine a modern nuclear power plant. The actual reactor room would easily fit into a medium-sized apartment building. However, the costs for cooling, fuel supply, and disposal, radiation safety requirements, and the actual generation of electrical energy expand it to the size of a small town. The situation is similar in semiconductor production. To manufacture a processor that is ultimately used in a smartwatch, you need production lines that fill an entire factory hall with a clean room atmosphere. Their preparation and supply alone devour millions, first in investments and later in ongoing operations.

It is no different in mechanical engineering. A high-performance packaging machine for medicine blister packs must meet high hygiene standards and extensive safety equipment must be provided for the operating personnel. Moreover, it incorporates numerous components and functions for quality assurance, making its complexity incomparable to early packaging technology machines.

This complexity extends to most modern processing machines, whether in the machine tool, plastics, or paper processing industries, to mention just a few.

### 1.2.2 The engineering process is changing

Until well into the 20th century, mechanical design engineers dominated the field, with design managers almost exclusively holding degrees in mechanics. Even after electronics-based automation technology emerged in the mid-1970s, the willingness to hand over mechanical expertise to chips and transistors was very slow to catch on. It wasn't until the end of the 20th century that the advantages of implementing new functions via software or using electric servo drives with electronic cam plates — offering both cost savings and flexibility over complex asymmetrical mechanical gearboxes without sacrificing reliability — began to be widely recognized.

The incipient fall in prices for electric drive technology and ever more powerful control components has led to a notable reduction in the mechanical aspects of design. This evolution has paved the way for new, innovative machine designs. As a result, the engineering components required in the development of a production system are shifting dramatically (Figure 1.5).

In the meantime, we no longer speak of mechanics and electronics as separate disciplines, but of mechatronics and mechatronic components with technological functions that are increasingly defined solely by software — and the trend is rising (Figure 1.6). In the future, and possibly already starting today, it will be irrelevant whether the head of development in a mechanical engineering company is a mechanical designer, an automation engineer, or a computer scientist, as long as a holistic view across all domains becomes the basic principle.

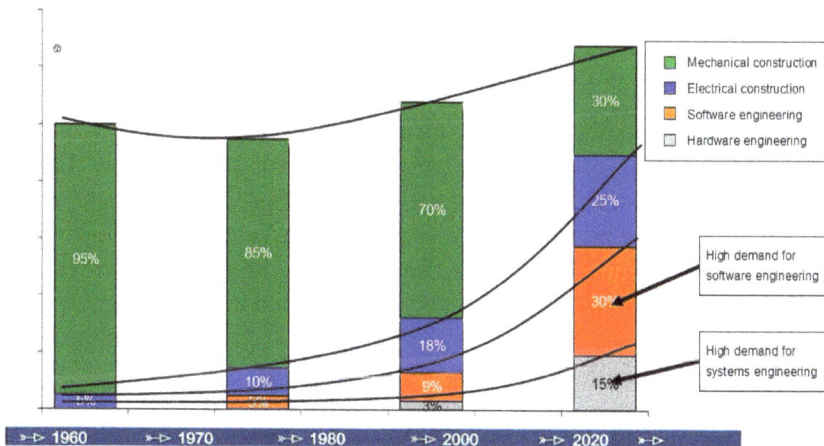

**Figure 1.5.** The share of engineering in mechanical and plant engineering is shifting toward automation and software development.

*Source*: ITQ GmbH.

**Mechatronics yesterday**

Mechatronics, a partial task

Drives
Safety technology
Machine construction

**MECHANICS**

Gearboxes

Design

MECHATRONICS

Infor-
matics

Hardware
**ELECTRONICS**
Field devices

**Mechatronics in the future**

Mechatronics, a holistic task

MECHATRONICS

Connection to ERP

**INFORMATICS**

Technology
control
software

Manufacturing
execution system

MECHA-
NICS

Security

Networks

Safety
technology

Drives    HMI

Condition
Monitoring

Hardware
**ELECTRONICS**
Field devices

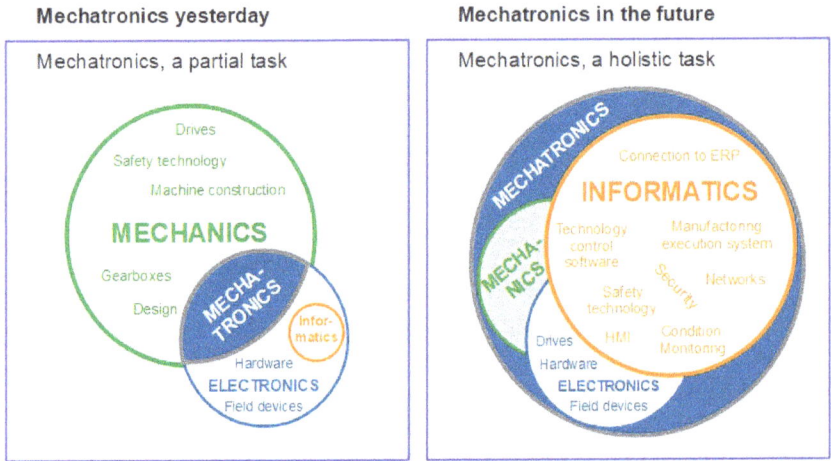

**Figure 1.6.** Mechatronics is currently shaped by mechanics. In the future, mechanics will only be a small component of a production system.

## 1.2.3 New requirements for production facilities

Following, the most important influences for this development and the resulting requirements for production systems will be examined in more detail. Conceptual approaches to automation technology solutions are presented later.

### 1.2.3.1 *Efficiency determines success*

Business success being primarily measured by monetary profit is in the very nature of our market economy. When Frederick Winslow Taylor presented his piece rate system to the world during his highly acclaimed lecture in 1895, the pursuit of efficiency was given a name: Taylorism. At its core, it is all about the organization of work, with the aim of minimizing unit times and therefore labor costs. The introduction of assembly line work in the early 20th century (2nd industrial revolution) was ultimately the logical continuation of Taylorism.

Automated systems such as automatic production machines or robots do little else than carry out programmed work steps as quickly as possible with maximum repeat accuracy (reproducibility). They carry out operations with the required precision that would be too much for humans, work without breaks, and do not become ill — a blessing for any company.

However, theory and reality often diverge. A high-performance machine, regardless of the industry, consists of thousands of individual components, lots of electronics, and several person-years of software development. It represents a highly complex system and can be prohibitively expensive. Every second of downtime not only means high costs but also leads to production delays and potential damage to the company's reputation if orders are not delivered on time. If a company has a just-in-time contract, it can be ruined by a prolonged production downtime.

Therefore, downtime, whether for conversion or due to faults, is a critically sensitive issue. Depending on the requirements, even enormous constructive and therefore economic effort can be justified to reduce them. Here are a few examples to illustrate this.

**Retooling and product conversion**
When a product changeover is due on a processing machine, the technology's complexity dictates the extent of required interventions. Tools need changing, material storage units loading, parameters adjusting, and much more. Test runs are then often carried out and changes are made to the settings after testing, measuring, and evaluating the prototype. This process consumes time, materials, and personnel, diminishing profitability. Moreover, this approach is entirely unsuitable for individualized products in batch size one.

Thus, it's a fundamental goal of all machine operators to minimize setup time as much as possible, with automation and mechatronics playing a key role. For example, adjustments and stops can be adjusted using electric motors, there are automated tool-changing systems, and all the necessary data is stored in a recipe management system integrated into the Human Machine Interface (HMI). HMI contains all the necessary data. By selecting the appropriate recipe, operators can initiate most of the conversion with a simple gesture on the touchscreen, allowing them to attend to other tasks until the automated process is complete. Downloading these parameters from a Manufacturing Execution System (MES) directly to the machine can further support the retooling, especially crucial for batch size one.

However, the number of actuators required in a processing machine can quickly become a significant cost factor. In a modern saddle stitching machine for the production of wire-stitched magazines and brochures, for example, over one hundred actuators are required. With the corresponding control electronics, cabling, and mechanics, the number of components — and ultimately the entire engineering effort — increases, significantly affecting the machine's pricing. Consequently, component selection is under intense cost scrutiny. After all, every dollar per actuator in the above-mentioned saddle stitching machine adds up to a factor of over 100 — a fact that is at the top of every price optimization agenda.

Developers are therefore well advised to compare the requirements of the product or process with the possibilities of the available mechanical and electrical components as precisely as possible right from the brainstorming stage. Every hour of careful research is time well spent on convincing arguments in budget meetings.

## Automation 4.0 Perspective

The retooling of a machine is just as automated as its actual production function.

**Maintenance and servicing**
Even the most modern production systems and the highest quality materials are not immune to wear and tear, necessitating regular maintenance and service. Production downtime due to a component failure or software error inevitably causes annoyance and frustration. When a necessary spare part is unavailable immediately, leading to procurement delays of days or weeks, and crucial orders are missed, tensions often run high. Such incidents frequently result in legal disputes. Irrespective of strict quality and supplier agreements, strategies and concepts are therefore required to rule out unforeseen failures as far as possible or at least minimize the resulting downtime.

One approach is the *"fire department" strategy* were, upon a machine's breakdown, a standby service team springs into action, fetching any required spare part from storage for immediate repair. This requires a more or less extensive stock of spare parts directly at the machine operator or service partner. Which components are kept in stock and which are not depends on their importance, replacement time, storage costs, shelf life, etc. However, such a warehouse and the necessary infrastructure, including maintenance personnel, tie up considerable capital and are therefore the most expensive and complex solution.

*Scheduled and preventive maintenance* entail the planned replacement of components deemed high-risk, alongside routine checks like tightening screws, inspecting cable harnesses, and performing oiling, cleaning, and lubrication. Recommendations from component and machine manufacturers and the maintenance staff's own experience are used to plan this type of maintenance. The resulting downtime is taken into account in the production plan, and spare parts and auxiliary materials are procured in advance.

The advantage of this procedure is the scheduled shutdown and the guaranteed maintenance of the production plant. However, components are sometimes replaced simply because the manufacturer recommends it, even though they would certainly have continued to work faultlessly, until a subsequent maintenance check. Even if the spare parts business is always a lucrative one, it should not be assumed here that maintenance intervals are specified by suppliers as shorter than necessary for reasons of profit. Rather, it is the supplier's duty to guarantee a service life, and in this respect, he is like the food manufacturer: He specifies a minimum shelf life, and depending on how the machine is operated, this may deviate more or less from the achievable one.

In this context, the operator acts cautiously, akin to how food retailers handle expired best-before dates, replacing components as a precautionary measure to avoid accusations of non-compliance with regulations and manufacturer guidelines.

Another disadvantage to this strategy is that if it turns out during the accompanying inspection that a component has to be replaced and procured again unexpectedly, delays are inevitable, and the delayed production start-up causes additional trouble.

As a result, *condition-based maintenance* is increasingly being practiced. This strategy involves systematic, automated analysis of machine data to monitor the state of (critical) components. For instance, the continuous current consumption of a motor can provide information on how the running characteristics of the driven mechanics are changing.[1] If the current increases steadily or abruptly compared to the long-term average value, this may be due to the onset of wear, lack of lubrication, or contamination. If, on the other hand, it drops, bearing play could increase and cause a failure. With the help of an additional ultrasonic sensor on the gearbox, changes can be detected in the continuously monitored

---

[1]When using frequency inverters or servo amplifiers, no additional measuring device is required to record the actual current value. With these devices, the value can simply be queried via a parameter call.

frequency band that indicate gearbox damage long before failure. Corresponding components are integrated into the automation system and the warnings are communicated directly to the operator on the machine display or to service personnel via the network. This information helps determine component longevity, replacement timing, and whether operational adjustments are necessary to avoid total failure until replacements are secured. Notably, *artificial intelligence (AI)* methods are increasingly utilized for this purpose.

The advantages are obvious: No or only minimal spare parts stocks and maximum utilization of the service life of highly stressed components relieve the maintenance budget, and unforeseen downtimes are largely avoided. While initial investment costs are higher, the savings from avoiding just one unplanned downtime can easily justify these expenses. Additionally, if the sensors also contribute to process optimization, the investment can pay off within months.

## Automation 4.0 Perspective

Intelligent machines themselves influence their maintenance.

**Resource consumption**

The significance of this topic extends beyond the rise of Industry 4.0. Resources such as energy and all materials and auxiliary materials required for production generate costs and have therefore always been the focus of production companies and, of course, their suppliers. However, with the Kyoto Protocol adopted in 1997 to limit and reduce greenhouse gas emissions, the issue of energy consumption has gained prominence.

Products of all kinds, including machines and systems, are increasingly being assessed according to their energy consumption, which must be shown, for example, in the case of light bulbs, household appliances, and much more through EU energy consumption labeling. This evaluation extends to calculating the energy consumption involved in manufacturing a product, allowing for an assessment of the production process's ecological footprint. As a result, machine and system manufacturers are increasingly confronted with the requirement to determine and document the actual consumption of resources per unit — the so-called gray energy.

New technological processes such as additive manufacturing or the use of carbon fibers in composite materials can also reduce material consumption and, as a result, the weight of a product, thus indirectly contributing to a reduction in energy consumption. Here, automation technology plays a crucial role, as it can greatly enhance the efficiency of various movement and dosing systems used in these generative processes, going beyond mere data collection and MES system integration.

For example, intelligent control algorithms can optimize the operation of drive, hydraulic, and temperature control systems to be more energy efficient. A typical example of this is the control of servo-electric pump drives in plastic injection molding machines. With this method, media flow and pressure settings can be optimized in such a way that the energy consumption of the hydraulic system in such machines can be reduced by around 50% [6].

## Automation 4.0 Perspective

Machines work in a resource-saving manner and can document actual consumption.

**Engineering effort**

The factors presented so far for increasing efficiency cover a wide range, but that is by no means all.

Increasing efficiency in mechanical and plant engineering also means reducing engineering effort. This challenge is not trivial, as heightened complexity typically leads to additional work. Often, there is insufficient time to address this complexity adequately, resulting in engineering tasks being deferred to the post-delivery stage at the customer's site, where they incur significantly higher costs and yield lower quality. To counteract this and prevent engineering costs from rising exorbitantly despite increasing complexity, the formation of mechatronic units and machine modules in the form of hardware and software is a common solution (Figure 1.7).

However, this also increases the requirements for automation, not only in terms of the scalability of hardware but also in the methods and tools for software development. Standard programming languages, structured methods for modularization, and systematic test methods are therefore among the most important features of an engineering tool.

Manufacturing effort

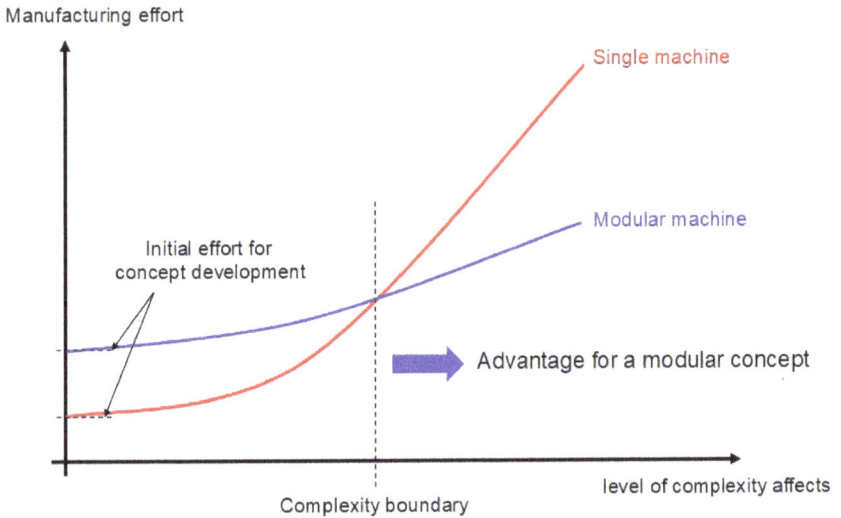

Figure shows two curves: "Single machine" (red) rising steeply, "Modular machine" (blue), with "Initial effort for concept development", "Advantage for a modular concept", "Complexity boundary", and "level of complexity affects" axis label.

**Figure 1.7.** With increasing complexity, a modular hardware and software concept brings clear advantages despite increased initial outlay.

*Source*: ITQ GmbH.

This book is specifically dedicated to this topic and the following chapters present suitable solutions.

## Automation 4.0 Perspective

Modular concepts enable the production of customized machines and systems at the touch of a button.

### 1.2.4 Service creates trust

The previous section discussed maintenance strategies, focusing on mechanical and hardware-oriented approaches. But what happens to the software?

From our personal experience with operating systems, applications, or smartphone apps, we're familiar with updates that introduce bug fixes and functional improvements. It is no different in modern machines and systems, and since the observations in Figure 1.6, we know that this will continue to grow. However, operators' service personnel are increasingly unable to maintain the software of a production system, either due to lack of capability or because product liability concerns prohibit them from doing so.

This challenge extends across the entire supply chain. Manufacturers of machines and components often rely on complex products from other suppliers, which they are neither capable of maintaining nor permitted to do so. They may be outright denied access, lack the necessary tools, special software, and training, or simply be overwhelmed by the complexity. Consequently, maintenance personnel often find themselves at a loss when faced with a network of systems, components, and their subcomponents.

However, automation engineers are not entirely innocent of creating these situations. The emphasis is on not entirely because they in particular have to submit to the trends mentioned in this chapter and sometimes have to conjure up the expected functions with more or less effort. Unfortunately, it is all too often the case that this is achieved by using hardware and software with the attribute "quick-and-dirty" that is inadequate or unsuitable due to sometimes excessive cost and deadline requirements. This is certainly not confidence-building.

So, what's the solution if the goal is to provide machine users with top-notch service, not least as a sales argument? When does maximum high-tech functionality have to be combined with minimized costs? When a costly production system needs to operate 24/7 without downtime? If the maintenance personnel should not only act as mediators between far too many interest groups but also *maintain* in the sense of *servicing*?

This calls for the automation engineer's expertise in responsible engineering. This involves a thorough analysis of requirements, careful selection of components, and sustainable software development, which naturally builds trust with customers and company management. The following chapters explore suitable methods for this purpose.

Irrespective of this, the topic of *software maintenance* is moving ever higher up the maintenance priority list. Remote maintenance technology was developed long before the Internet became widespread. Software developers or service personnel have been able to access the control system of a machine via telephone lines and modems for decades, nowadays via the Internet and secure data connections, no matter where they are in the world. This allows for quick and precise troubleshooting guidance or dispatching technicians with the correct spare parts.

With the possibilities of the Internet, remote maintenance and diagnostics can be realized with a completely new quality and quantity. Machines and sometimes all of their intelligent components can be defined as cyber-physical objects in the *Internet of Things*. They can communicate their status independently, provide product and quality data, or communicate with authorized persons at different hierarchical levels. The only hardware required is an Ethernet connection, either in the form of a cable or WLAN. The rest is managed through software.

However, this technology is not without its risks, as impressively evidenced by viruses like Stuxnet [7]. It is vital to invest in additional data security infrastructure for both hardware and software.

Another problem is the variety of protocols used by different components, machines, and devices. In addition to many bus systems, which are more or less suitable for real-time capability and data throughput, there are other industry-specific and sometimes completely unique protocol formats that are used as a transport medium.

We address this issue in Section 5.4 and outline possible solutions.

> ### Automation 4.0 Perspective
>
> Machines are an object in the Industrial Internet of Things (IIoT) via standardized interfaces and can therefore be accessed worldwide.

### 1.2.5  Quality is unconditional

Have you ever been frustrated by the poor quality of a product you purchased? Did you share your disappointment with friends or post a negative review on social media?

Negative reports about product defects spread faster than wildfire on the Internet across all borders, so it is not surprising that the topic of quality assurance is attracting great attention. Spectacular claims for damages in the multi-million range further fuel this urgency.

The issue of quality assurance is not simply about manufacturing a product without defects. Instead, manufacturers are keenly interested in demonstrating that their production processes are impeccable. To achieve this, extensive data is collected during manufacturing and linked to each product ID in a database. Furthermore, the serial numbers of each individual sub-component are stored, as are the names of the employees who outsourced, transported, and finally assembled them. This also includes all measurement and test logs and, finally, the total energy consumption from each process step.

Do you think this sounds excessive? It's actually quite the contrary.

For instance, the U.S. Food & Drug Administration has introduced a set of regulations into American law that obliges every drug or food manufacturer and their suppliers to keep all relevant quality data for each individual pack of medication and to produce it on request.[2] If there is a problem with one or more patients, it is possible to compare the data of the actual product with the original data from the approval procedure, detect possible causes, and take the resulting measures.

Almost identical processes are used in automotive, aircraft, and rail vehicle construction. Each individual component is precisely recorded with all its individual components. Safety-relevant production data such as the tightening torque of an important screw connection or the pressing pressure of an adhesive seam are automatically recorded and documented. If, for example, a defective component is identified as the cause of an accident, its history can be used to precisely identify whether there is a series defect and whether other vehicles may also be affected.

Thus, automation technology is expected to operate flawlessly, necessitating additional sensors like barcode scanners, image recognition systems, and ample storage capacity for data collection, transmission, and/or storage. In addition, this data must be recorded in real time, i.e. synchronized with the process. This can put an additional strain on the performance of a control system and must therefore be taken into account during planning. Generally, decentralized automation systems can manage performance loads more effectively than centralized configurations. With changing data storage needs, a central hub can quickly become a bottleneck, whereas decentralized systems benefit from an appropriate real-time communication system.

Sections 5.4.3 and 5.4.5 deal with this topic in more detail.

## Automation 4.0 Perspective

Production systems ensure quality and efficiency through comprehensive data acquisition and communication.

---

[2] Further information is available at http://www.fda.gov.

## 1.2.6 Adaptability makes you fit for the future

In Section 1.1, we discussed the challenges of ever-shorter innovation cycles and the trend toward the individualization of products. Now, we explore the implications for machines and systems. What happens if the relay can no longer be produced in millions of units because the technology used becomes obsolete faster than expected? Or when even a customizable product falls out of demand? If a material required for a certain process is suddenly declared harmful to health and banned?

Permanent renewal has been a side effect of industrialization since its beginnings, but the few examples show that production facilities will no longer remain unchanged for the usual operating times. Conversions, expansions, and modernizations will be required at much shorter intervals. Added to this is the already mentioned increase in complexity, which, without appropriate design concepts, will quickly cause the cost of such interventions to explode.

Therefore, machines and systems need to be extremely flexible and interchangeable over their entire life cycle. This starts with the production process, incorporating customer-specific option packages, and continues in actual use on site. Ultimately, the aim is to be able to optionally select and configure certain technological functions for the production of the respective end product. The mechatronic approach of modularization is particularly well suited for this purpose.

Several goals are being pursued here: First, a modular machine design allows production costs to be reduced through largely customer-neutral production capacity utilization with a series character. This is achieved by personalizing the individual customer machine as late as possible. Second, it can extend the lifespan of a production facility by allowing for the case-by-case replacement of obsolete machine modules with modern ones. This enables operators to upgrade their systems incrementally, in line with technological trends, without the need for entirely new systems. Consequently, when designing a production plant, it's crucial to consider from the beginning how it can adapt to new tasks without rendering the entire investment obsolete.

As can be seen, modular machine and system concepts are the key to meeting most requirements in both mechanical design and automation technology and are therefore dealt with in more detail in the following.

> **Automation 4.0 Perspective**
>
> With the modularization of production facilities and the entire associated design process, individual machines and systems are created at the cost of series products.

### 1.2.7 Safety is a must

Every technical device can pose a danger to people, the environment, and, in principle, everything. A needle can cause a prick, contact with a live electrical part may result in burns or even death, and an accident in a nuclear power plant could render entire regions uninhabitable for centuries, affecting thousands of individuals. Similarly, machines and systems carry inherent risks. Operators must be protected from collisions with moving machine parts, such as high-speed drives, press beams, or robots, and the safety concept of machinery and systems must also protect the environment from the escape of toxic substances or explosions. Moreover, high-quality products need to be safeguarded against potentially irreversible damage due to machine malfunctions or electrical supply failures.

In order to determine the risk of human injury, a continuous risk assessment must be carried out throughout the entire engineering process in accordance with standards IEC 61508 and IEC 61511, primarily for systems, and ISO 13849 for processing machines. Figure 1.8 shows how the risk of a controlled device can be gradually reduced to an acceptable risk and beyond.

This begins with external and non-electrical measures (access restrictions, protective grilles, and covers) and extends to electrical engineering measures (e.g. simple interlocking with centralized shutdown by switching off) and automation technology (safety control systems). Numerous publications (e.g. [8–10] and many more) deal with these measures from different perspectives. For this reason, this book does not discuss the risk assessment and application of standards and regulations in detail. The same applies to measures relating to the protection of the environment.

However, there is an increasing number of safety requirements for processing machine construction that are not or only subordinately part of

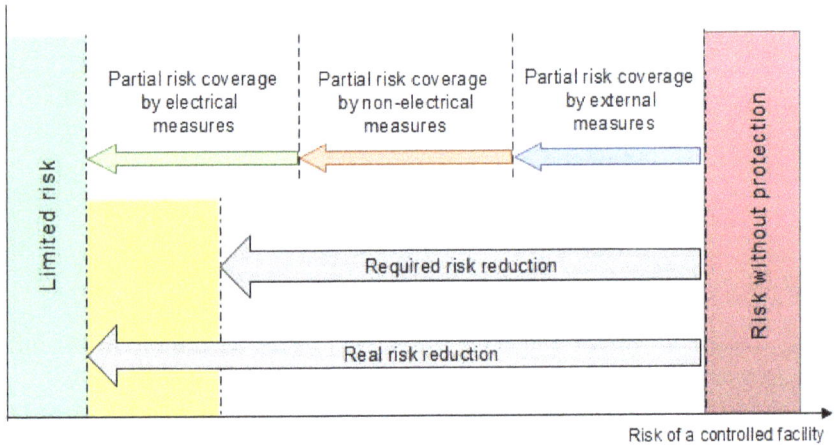

**Figure 1.8.**   The risk of a controlled device must be reduced to an acceptable level by a combination of electrical and non-electrical measures.

the standards but which are becoming increasingly important for the design of an automation system.

**Protection of production facilities**

Machines and systems are cost-intensive production assets, and as we learned in Section 1.2, service failures and the associated downtimes must be avoided. However, with the advent of new technologies, functions are increasingly being integrated that can cause damage to mechanical components if they malfunction.

A typical example of this is coupled servo axes, for example, for embossing with a rotating upper and lower beam (Figure 1.9). In this application, the speeds of the upper and lower beams ($\omega_1$, $\omega_2$) must be synchronized so that the two beams are angularly synchronous and do not collide.

The mechanical gears previously used for this task had the decisive advantage that — once correctly adjusted — they inevitably worked collision-free due to the rigid coupling. The disadvantage of this mechanical design, however, is that when the product format is changed (conversion from embossing distance $a_1$ to $a_2$), the corresponding gearboxes must be recalculated, manufactured, and installed — a very time-consuming and costly process. As the number of product variants increases (see megatrends in individualization), this becomes a prohibitive drawback for such designs.

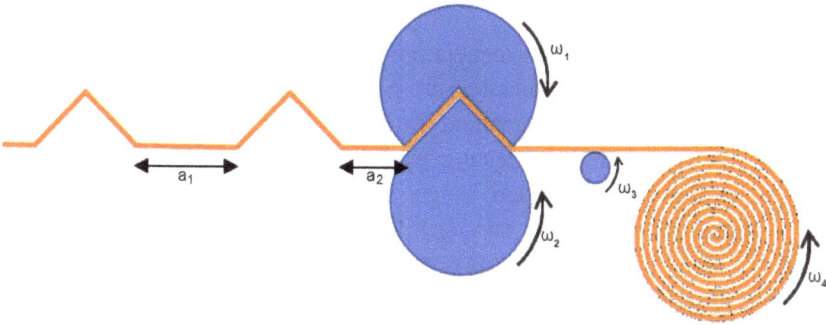

**Figure 1.9.** Diagram of an embossing device: The variable distances shown can only be realized with separate servo drives.

The solution lies in modern servo technology, where the two tools are driven separately, and the movement functions can be freely configured. This means that, as shown in Figure 1.9 the distance *a* can be changed from one embossing to the next without mechanical adjustment, simply by changing the parameters, even while the movement is running.

However, this unbeatable advantage entails the risk of the tools becoming wedged and irreparably destroyed in the event of incorrect parameterization or a hardware or software error. In addition, the bearings and other machine components involved can also be affected.

Even when industrial robots are used, hazardous system parts within the movement space must be protected against collision. It is therefore necessary to monitor the movements of the axes within their working space limits for safety reasons. In addition, the drive or the entire system can be shut down, for example, if an impermissible increase in torque occurs in the event of a collision. In addition, the relative rotational movement of the axes can be monitored and, if a permissible deviation is exceeded, the drives can also be stopped.[3] Although this results in an unplanned machine stoppage, the repercussions are significantly less severe than in the initial scenario.

---

[3] For safe stopping, suitable drive systems are required that enable coordinated movement to a standstill in this application even in the event of a power failure. In practice, this is achieved using local electrical or mechanical energy storage systems.

### Protection of products

An application for quality inspection in the semiconductor industry is described in [11]. Contact needles are positioned on wafers with accuracy in the micrometer range and held with a defined contact pressure. The subsequent measuring and testing processes involve analyzing currents sometimes in the femtoampere range.

In this application, the risk is that a single misplacement of a few micrometers can render the entire wafer or a pin board with over 10,000 contacts unusable. This would cause damage amounting to several thousand dollars and an enormous loss of image. If a prototype developed over a long period by an entire institute were affected, the damages would be even greater.

This example shows that, in addition to people and production equipment, products must also be protected against damage caused by malfunctions. However, there are neither laws nor regulations for this, and it is up to the machine and system manufacturer alone to decide how far they want to go with this effort or have to realize it on behalf of the customer. The value or consequential damage of a product that may become unusable is usually the decision criterion because it makes a difference whether an individual photo book worth twenty dollars or the aforementioned wafer can be destroyed. From the point of view of automation technology, however, it usually makes no difference who or what needs to be protected. Depending on the type and scope of the safety requirement, the same methods, devices, and components can be used.

### Prevention of manipulation of protective devices

What is the value of the finest safety devices if they can be disabled by unauthorized tampering?

Unfortunately, such occurrences are not rare, as demonstrated by a study conducted by the German Federation of Statutory Accident Insurance Institutions [12]. In order to find out the reasons for the manipulation of protective devices on machines, 940 surveys were carried out among occupational safety specialists, with the majority of machines built before 1995, i.e. more modern machines, being examined. The results are both sobering and alarming.

The study indicated that nearly 37% of protective devices on machinery are permanently or temporarily manipulated, with around half of these

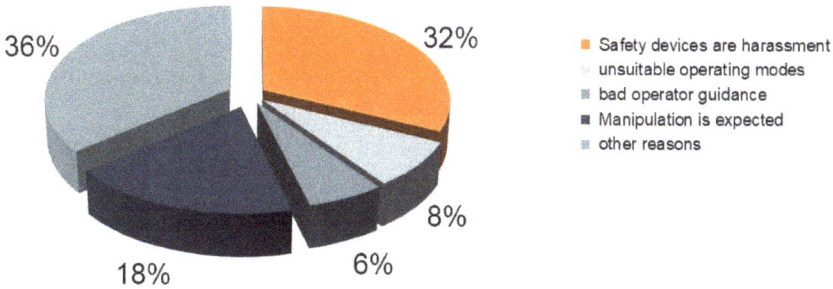

**Figure 1.10.** Eighty percent of tampering with protective devices is justified by the desire for greater effectiveness.

*Source*: Ref. [12].

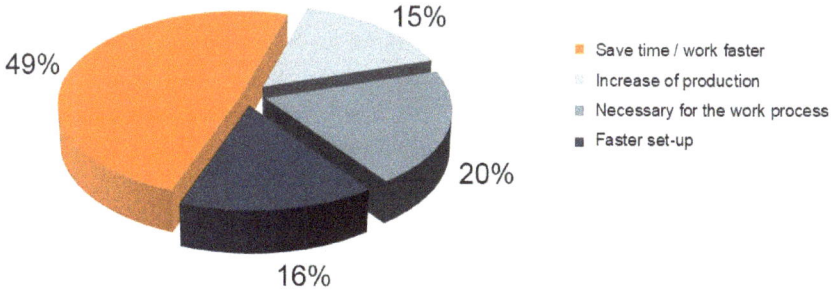

**Figure 1.11.** Almost half of the manipulations detected were due to inappropriate procedures or alleged harassment.

*Source*: Ref. [12].

leading to accidents. Furthermore, approximately 41% of workplace accidents on machinery occur during "operation and control". The question therefore arises: Why is this the case? Why do operators override elaborate protective equipment installed solely for their personal protection?

Like Figures 1.10 and 1.11 illustrate, these are often seen as harassment, appear to hinder the work process, or worsen workflows. In the category "Reasons for manipulation", almost half of the respondents refer to the optimization of the work process in terms of time. However, protective devices are also manipulated to speed up setup, i.e. to reduce setup times.

It is also worrying that 14% can be attributed to poor engineering, as unsuitable operating modes and poor operator guidance are nothing other than poor design work. In extreme cases, this can even lead to the opinion that alleged "over-automation" hinders flexible working [13].

The following example [14] demonstrates an alternative approach: At a manufacturer of machines for packaging chocolate bars, dishwasher tabs, and the like, complex special gearboxes were to be replaced by several coupled servo drives in a machine series. In addition, the designers — similar to the example in Figure 1.9 — saw considerable advantages in the realization of asymmetrical motion laws that were mechanically not possible.

However, these specific axis configurations introduced potential hazards during setup operations. If the axis assembly had previously been moved into the corresponding setup positions by means of a handwheel, the operator could react immediately in the event of painful contact with a blade, for example, and prevent worse. Safety functions were therefore required in the drives to guarantee the same level of safety for the operator as before in the event of a fault. The project was shelved for a long time, as these were not yet available at the time. The project resumed only after the introduction of the "safe limited torque" function in the servo drives, coinciding with the trend toward ever-smaller batch sizes. After the first successful tests, in which the torque of the critical drives was reduced to a safe level and safely monitored in setup mode, another problem arose — the method of operation.

To avoid overwhelming or alienating existing customers with new operational philosophies, it was crucial that the operator interface retain the familiar haptics of previous machines. An automation solution was found whose simplicity surprised even the most conservative mechanical designer. The existing automation concept only had to be supplemented by an electronic handwheel and a few software functions (Figure 1.12).

The operator can now get much closer to the action by simultaneously pressing a safety enabling switch and the usual turning movement on an electronic handwheel, which could even be mounted on a mobile control panel. The implementation of electronic safety functions means there is only a marginally higher residual risk for the operator, a factor deemed acceptable by the employers' liability insurance association for approving this design solution.

This example illustrates how safety-related equipment can be integrated in such a way that operators are discouraged from attempting to pass it. In this case, an additional advantage has even been generated, as the designers have integrated a function that allows critical setup positions

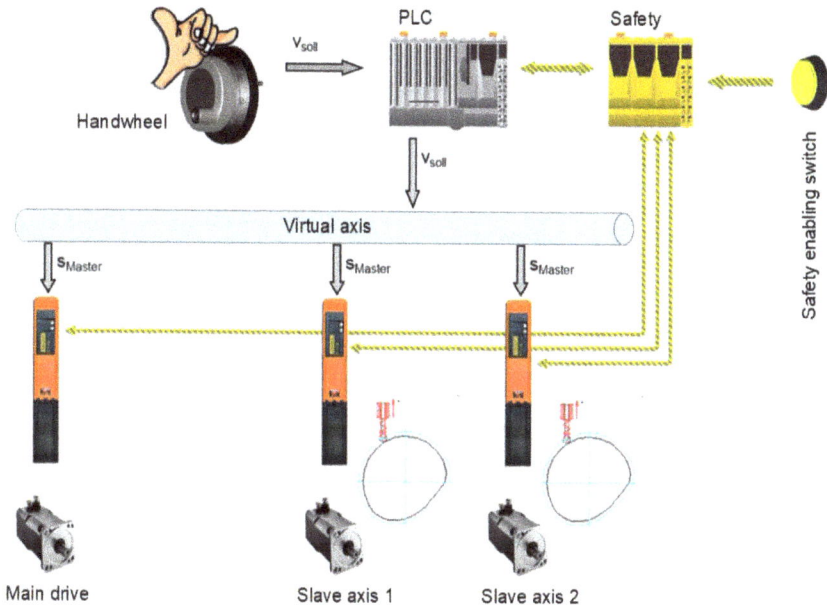

**Figure 1.12.**   Integrated safety technology and safe motion functions enable safe setup operation to be realized.

to be approached automatically and safely. This simplifies the process and reduces the conversion time, which ultimately results in a unique selling point — a classic win-win situation.

In another case, the conversion time on a saddle stitcher was also to be drastically reduced. Customer surveys revealed that the operational process could accommodate multiple individuals working on the setup simultaneously. However, with the technology used, this could only be achieved with an unacceptably high residual risk, and so this requirement was repeatedly rejected. Operators resorted to bypassing protective cover monitoring switches and even leaving them open during production until a near-miss incident prompted a reevaluation.

Here too, new Safe Motion Functions help. In a revised generation of machines, the Safe Torque Off (STO) and Safe Limited Speed (SLS) functions are activated in the drives when switching to setup mode. At the same time, each operator can be equipped with their own hand-held pendant station. This makes it possible for any number of setters to move individual drives at the same time and thus make their settings without creating an additional safety risk for the others involved. Manipulation is therefore no longer necessary, and a customer requirement has been met.

These examples impressively demonstrate that the consistent use of new control technology options not only eliminates the motivation to manipulate the safety devices but can also generate new usage properties.

---

### Automation 4.0 Perspective

Safety and flexibility are not mutually exclusive, but when cleverly combined they generate additional customer benefits.

---

### 1.2.8 Turning new technologies into success

The examples in the previous section would have also fit perfectly in this section, as safety technology integrated into automation systems has made an enormous technological leap in recent years. Thus, it's no surprise that many new technological capabilities are closely linked to this progress. As the following examples show, it is not just about reducing production costs by saving time and resources, even if these aspects will never lose their importance.

**New production processes**
The procedure described in Section 1.2.1 inevitably requires a wide range of new production processes. One example is the processing of carbon fibers for the production of high strength yet extremely lightweight body components. These are needed, among other things, to reduce the weight and thus the energy consumption of cars and aircraft [2]. Furthermore, resource-saving materials made from reusable and/or natural raw materials are increasingly replacing those made from metal or plastic. However, their processing technologies sometimes differ significantly from previous methods. Whereas metal body parts are bent, welded, or riveted, carbon components are knitted, baked, and glued.

Naturally, these novel processing technologies also require more sophisticated automation technology in hardware and software. For example, carbon fiber processing produces extremely fine carbon dust, which is downright deadly for electronic components due to its conductivity. Consequently, hardware solutions with enhanced protection levels are necessary, particularly for field devices and decentralized components within the production area.

Furthermore, different processes are required for pressing and forming natural materials than for metal. Gentle immersion in the material, parameterizable dwell cycles, and adapted pressing forces require new functions in the presses. The conventional eccentric press is unsuitable for such processes. One solution is the use of direct servo drives with safe drive functions. This allows finely parameterizable movement and pressing force profiles to be run and, as a welcome side effect, saves energy.

**Safety, cooperating, and collaborative robots**
When talking about robots, most people immediately think of images of human-like figures or large robot farms in car manufacturing. We are familiar with the former from science fiction films, as fighting machines or as peaceful and almost omniscient assistants. In the meantime, the latter vision has also arrived in the real world. Whether as a service robot in (Figure 1.13) elderly care or as a heat-resistant firefighter: A survey

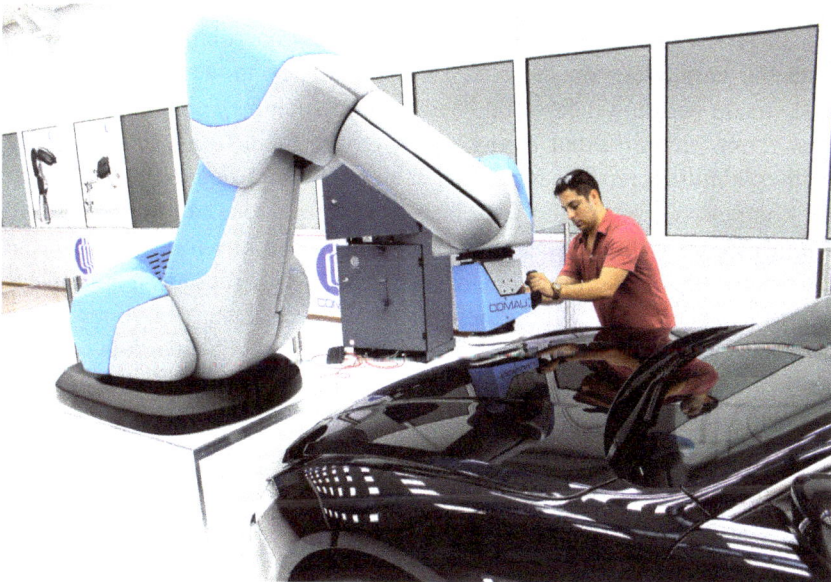

**Figure 1.13.**   Humans and robots work as a team: Thanks to its touch-sensitive outer skin and an extremely responsive safety control system, the robot poses no danger.

*Source*: Comau S. p. A.

published by the German Federal Ministry of Education and Research (BMBF) revealed that 83% of Germans could envision using a service robot at home, enabling them to age in place [15].

Realizing this vision involves overcoming several challenges, including efficient, reliable, and secure communication, interaction, and collaboration between humans, autonomous systems, and their environment. In order to be accepted by humans, these systems must adapt to their communication behavior and there must be ways of transferring control from and to humans. As a simple emergency stop is not sufficient in the event of a problem, the handover must be regulated, transparent, and appropriate to the situation [16].

In the production environment, the scenario is no different. Robots are already being released from their cages and work together with humans on a task. This technology is basically about combining the advantages of robots with those of humans in a way to create or optimize new work processes. With their fine motor skills and experience, humans are the key to high flexibility, while robots take on ergonomically demanding tasks, for example, and thus guarantee maximum repeat accuracy and quality [17].

In AUDI production, for example, inspection robots work directly with the employees on the assembly lines. They move various camera systems on pivoting measuring heads, with which they can view the door gaps from very different angles and measure them to an accuracy of a tenth of a millimeter. The measured values are forwarded to the body shop for comparison, where the pre-assembly of the doors can be carried out with this data in higher quality. At the same time, the robots relieve the colleagues on the assembly line of a stressful task: Previously, employees had to squat or get down on their knees to check the lower zones of the door gap using a hand-held measuring device. Now their work is concentrated on areas that are ergonomically safe: the hood and the luggage compartment lid [18].

However, if humans and robots share a common workspace or movement area, physical contact between the two is often unavoidable or even intentional. This generally results in a high-risk potential. Injury to humans by the robot must therefore be safely avoided. To mitigate this, standards such as DIN EN ISO 10218 and ISO/TS 15066 [19] are applicable. Detailed requirements for all task-dependent safety distances, maximum travel speeds, and workspaces as well as their monitoring can be derived from these standards, and the requirements for the safety technology used are correspondingly high. Only the systems for safe sensor technology, integrated control, and actuator technology that have reached market maturity in recent years bring the required sensitivity and reaction speed into an acceptable safety range for collaborative robots. In this way, new safety technologies are acting as a catalyst for many future applications, especially in the field of robotics.

### Reduction of safety distances

Press brakes, cutting machines, and many other machines require direct interaction during normal operations, placing operators close to hazardous zones. Despite this, increasing process speed and performance remain priorities for manufacturers. However, safety regulations, particularly in these applications, often constrain developers.

Press brakes place some of the highest demands on safety technology. In these applications, the working area underneath the upper beam is permanently monitored using optical systems specially developed for this application. If this area is violated, the downward movement must be stopped as quickly as possible. At a standard process speed of 200 mm/s, however, this results in a relatively long stopping distance, which is determined by the mechanics used, the drive, and the reaction time of the entire safety control system. In practice, the speed is therefore reduced to a safe 10 mm/s shortly before the clamping point is reached — the moment at which a hazard is present. Its position results from various factors in the safety analysis and is predetermined. The challenge lies in positioning the breaking point so close to the clamping point that the machine can operate at maximum speed for as long as possible, thus enhancing throughput and reducing operator strain.

This requirement can only be met more and more effectively with modern safe control technology. The reaction times of the safety application are particularly important here, as they determine how close an operator can get to the source of danger without compromising safety.

The aforementioned aspects are intended to demonstrate the diverse potential that can be tapped into using intelligent safety technology. We therefore look at the topic of safety in more detail in Section 5.3.

> ### Automation 4.0 Perspective
>
> New technologies create unique selling points for manufacturers and customers.

## 1.2.9  Digital production

Even if the topic of digital production in the context of Industry 4.0 has already been discussed several times, it deserves special consideration once again. Just like automotive production, the end-to-end value creation of a photo book is already one of the beacons of digital production. This case particularly illustrates how the entire business ecosystem — from development and production to the organization of just-in-time logistics — can be fully automated.

What has also already been discussed, but is often overlooked, is that the production of machines and systems is also undergoing change. If series machines mutate into individualized special machines and are not allowed to cost more, then their manufacturers face exactly the same challenges. In order to master these, new thinking is required as early as the design process, alongside the adaption of new technologies. This includes coordinated development tools, early simulation, the integration of external expertise, and a reorganization of internal processes.

**Increase in overall equipment effectiveness**
Digital production also means increasing the *Overall Equipment Effectiveness (OEE)* of individual machines through to the entire production plant, intelligent maintenance strategies, automated plant optimization, and flexible process design. To this end, lots of data is collected, analyzed, and evaluated using a wide variety of methods. The term *big data* has been coined to describe the flood of data generated in this way. However, the question arises as follows: How can machine operators and plant managers feasibly oversee and, more importantly, interpret terabytes of production data? Moreover, which data will prove to be genuinely useful in an emergency or a crisis?

Without intelligent analysis tools, particularly AI methods, navigating this data deluge is impractical. The main aim here is to filter out the data relevant to the moment, the condition, or the individual product, separating

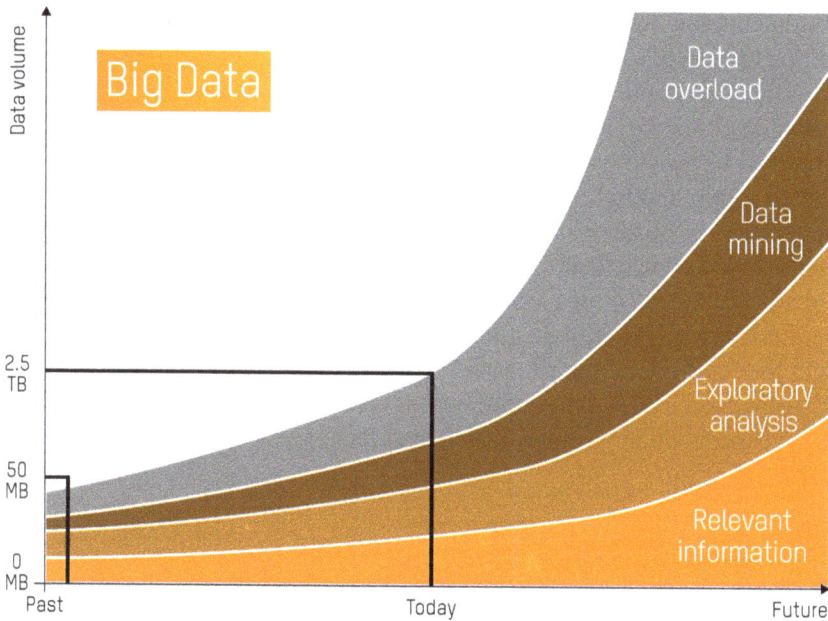

**Figure 1.14.**   Only a part of the data generated is relevant — and the trend is downward.

*Source*: B&R Industrial Automation GmbH.

the wheat from the chaff, so to speak. Studies of production data from various systems with process control technology have shown that only 10–20% of the data collected is relevant for process control, and only 1% of the data generated in production is used effectively (Figure 1.14), [20]. This includes current process states (e.g. temperatures, pressures, and speeds) as well as quality data and status data of the system components. However, statistical methods (explorative analysis, data mining, etc.) can be used to gain further insights into plant status from around 50% of the data pool. AI methods in particular play an increasingly important role here. As things stand today, the rest is still unused data that only costs storage capacity and may one day be analyzed using technologies yet unknown.

System providers are thus investing significant resources into the development of sophisticated assistance systems. These systems aim not only to enhance the controllability of machinery but also to improve product quality and service. Numerous publications and product documentation deal with this subject so that this topic is only mentioned in this section [21–25].

But what implications does this have for manufacturers of production systems? Looking ahead, it's likely that every manufacturer will be expected to provide all relevant data from each specific production step (for example, the tightening torque of a safety-critical screw connection or a photo of blister packaging filled with tablets), as well as from the machine or system itself, through standardized interfaces. Sensors are already able to send measured values directly to a cloud. There will also be a conveyor belt with an Ethernet connection, which will provide the data for condition monitoring of important components.

Manufacturers are therefore well advised to consider the requirements for data acquisition, pre-processing, and local storage early on in an automation concept.

## Automation 4.0 Perspective

Machines and systems provide data to increase your own and overall effectiveness.

**Interoperability in the production environment[4]**
Digital production fundamentally entails the networking of all involved production systems to optimize processes and unlock new value creation potential, extending even beyond the confines of a single company. Only this approach leads to greater sovereignty and sustainability, whereby *sovereignty* means that all market players are free to define and shape their individual business models right up to the individual purchase decision. *Sustainability* is fostered by leveraging digital technologies to embed sustainability goals and principles within the production process firmly. For instance, data sharing can reduce $CO_2$ emissions, conserve resources, and enhance operational efficiency by making vital information accessible, such as on the consumption of raw materials or emissions throughout the entire value chain.

Achieving this goal requires a well-developed infrastructure, data security, and an appropriate legal framework, as well as excellent training

---

[4]The explanations are based on [27, 28].

and further education opportunities. However, the most important prerequisite is a high degree of interoperability. Systems, devices, or applications can only communicate and network autonomously when this is achieved [27].

However, technical assets[5] are often still far from what simple devices have long offered in the office environment: Plug & Play from delivery to operation. While technologies familiar to everyday office life such as Ethernet, USB, and driver architectures already offer universal and manufacturer-independent communication standards, such solutions have hardly been widespread in industry to date. Instead, a wide range of proprietary and standardized solutions have been established for various applications. However, true interoperability in the industry requires much deeper software integration for assets and systems.

For example, the current situation often requires laborious efforts to connect an extraction robot on a plastic injection molding machine to the machine control system to synchronize workflows, sometimes relying on outdated parallel wiring due to the lack of functionally compatible serial connections on both sides — an anachronism in the digital age! Almost all companies involved are therefore looking for solutions for genuine interoperability and, in the view of the associations, proprietary and closed interoperability solutions are hardly sustainable in the long term [28].

All players in the Industry 4.0 platform environment are literally doing pioneering work here. As can be seen from the current progress report of the Industry 4.0 platform [27], various working groups are only concerned with comprehensively advancing this interoperability along the entire value chain. Several standardized interoperability solutions already exist, such as OPC UA,[6] AutomationML,[7] and the asset administration shell.[8] Reference [28] shows a target figure that illustrates how these

---

[5] For more information, see Section 2.2.4.2.

[6] See Section 5.4.3.

[7] AutomationML (AML) is a standardized, flexible, object-oriented data modeling language developed by AutomationML e.V. and a file format for storing and iteratively exchange of object models. AML was specially developed for repeated data exchange in engineering and enables consistent, digital, lossless data exchange in engineering tool chains [27]. Due to this categorization, AML is not discussed further in this book.

[8] See Section 2.2.4.1.

technologies together create a modern common interoperability landscape based on the RAMI4.0 reference architecture model.[9]

The following open standards are named for use:

(1) *AutomationML* for the asset development lifecycle as well as planning the productive use of the asset including all content and for communication in all networks.[10]

(2) *OPC UA* for the life cycle productive use and maintenance of the asset including all content (e.g. operating data) and for communication in all networks.[11]

(3) Across all life cycles, the *administration shell (AAS)* is recommended for communication in the Connected World network,[12] including all content that can be assigned to the manufactured product (e.g. digital type plate or $CO_2$ footprint).

(4) As the *OPC UA* communication standard and the function of the asset administration shell are particularly important for the automation of production systems, we go into these topics in more detail in separate chapters.

---

⮎  **Automation 4.0 Perspective**

Machines and systems interact on the basis of standardized methods and protocols according to the Plug & Play principle.

---

## 1.3  Conclusions

At this juncture, some readers may wonder whether it's feasible to meet all these outlined requirements. Our response is unequivocally positive: Yes, of course! A primary goal of this book is to chart a course toward achieving this.

We aim to demonstrate that even a single machine manufacturer can offer products that satisfy all the mentioned criteria at a reasonable cost,

---

[9]See Section 2.2.4.4.

[10]Except for Connected World.

[11]Except for Connected World.

[12]The Connected World network is the manufacturer's value chain and the network from which the manufacturer's customers have access to product information, i.e. mostly the Internet [27].

thereby securing market advantages. Manufacturers can leverage innovative technologies from their suppliers, as the solution spectrum is broad and nearly every component manufacturer has these requirements on their agenda. Automation technology providers, in particular, recognize their critical role in this ecosystem and have comprehensive technology packages in their portfolios to address these challenges, alongside the requisite hardware and software products. But is it sufficient to merely develop new functions, select components wisely, and establish new production facilities using existing concepts and methods?

Here too, the authors' answer is clear: No, of course not! This is precisely what leads to ever-increasing complexity, one-offs, and inflexible machines and systems with precisely the negative effects discussed above. In addition, dependencies on suppliers arise, which sometimes massively contradict the company's own requirements for innovation. Instead, it is necessary to face up to market trends, eliminate technological dependencies as far as possible, and develop and consistently pursue suitable strategies.

At this point, we return to the modularisation approach mentioned in Section 1.2.3.1 because it is the answer to increasing complexity and many other aspects. It is important to define *the market-compliant diversity of the product range and to design products in a modular way, just like services. The same applies to the design of the organization and production* [26]. If we focus mainly on the automation-related aspects of the modularization of machines and systems in the following, this does not mean that these are sufficient for a successful modularization concept. Consequently, [26] goes on to say that *the mere modularization of products and services is no longer sufficient in today's competitive environment. Many manufacturing companies have therefore also started to modularize production in order to reduce the investment and operating costs of their production facilities.* However, this is also precisely the entry into digital production so that only an Industry 4.0-compliant modularization approach can be effective for the future.

As described in Section 1.2.2, the functionality of a production facility will increasingly be represented by software in the future. And because modularization is an established method for reducing complexity in software engineering, it makes sense to fall back on the analysis and design methods commonly used there. The object-oriented approach is particularly important here, as it helps us to describe modular architectures. In addition, the resulting models can be incorporated seamlessly into the specifications for software engineering through to the design of the operating concept and documentation. Object-oriented models also help increase OEE for direct production planning, materials management, and controlling. We therefore begin with the selection of some technical terms, look at suitable methods for modularizing automated machines and systems, and apply these findings in case studies.

However, the transition from a complex individual machine to a modular series machine also generates various new problem areas that must be considered in a viable concept from the outset. For example, suitable joining methods (plugging and screwing instead of welding or gluing) must be introduced in the mechanical design for module assembly, the media supply and disposal (suction and blowing air, hydraulics, and electrics) must be adapted, and, last but not least, suitable handling criteria must be taken into account in order to be able to transport and use modules. It is also important to design the control technology in a modular way, provide it with a field bus, adapt the safety concept, and still make it easier to operate. All of this can in turn generate restrictions for real-time capability. This (certainly incomplete) illustration is intended to provide an example of the issues that designers have to face throughout the entire engineering process. As far as the automation part is concerned, we deal with the most important aspects in detail and point out possible solutions.

Thus, our discussions on project planning for modular machines and systems outline a roadmap for creating individualized series machines that can thrive in a digitalized production environment.

# References

[1]    Schröder, D.: What doesn't fit is made to fit, in: Brandeins, No. 07/2015.

[2]    BMW, A.G. and Kranz, U.: i3/i8 — Requirements for the supply industry's ability to innovate, in: Tagungsband ACOD 2014, Leipzig 2014.

[3]    Bendel, O.: Gabler Wirtschaftslexikon, Berlin 2015.

[4]    Verstraete in mold labels: What is IML?, available online at: http://www.verstraete-iml.com/de/was-ist-iml, 2016, last accessed: 11.07.2016.

[5]    Siemens, AG: Siemens achieves breakthrough with gas turbine blades from the 3D printer (press release PR2017020154PGDE dated February 6, 2017), available online at: https://www.siemens.com/press/pool/de/pressemitteilungen/2017/power-gas/PR2017020154PGDE.pdf, last accessed: 18.03.2018.

[6]   B&R Industrie-Elektronik GmbH: The revolution in hydraulics (B&R Industrie-Elektronik customer magazine) in: Automotion, No. 09/2013.

[7]   Holland, M.: Stuxnet allegedly part of a larger attack on Iran's critical infrastructure, in: Heise online 2006, available online at: https://www.heise.de/security/meldung/Stuxnet-angeblich-Teil-eines-groesseren-Angriffs-auf-kritische-Infrastruktur-des-Iran-3104957.html, last accessed: 25.07.2018.

[8]   Henke, W. (TÜV SÜD Rail GmbH): Functional safety in the manufacturing industry, TÜV SÜD Rail GmbH, B&R Safety Workshop, Bad Homburg 2010.

[9]   Siemens, A.G.: European Machinery Directive — simply implemented. Functional Safety of Machines and Systems, Erlangen 2012.

[10]  Heinke, B.: Safety-related control systems. Implementation and application of EN ISO 13849 — 1, published by Phoenix Contact, Bochum 2009.

[11]  B&R Industrie-Elektronik GmbH: Wenn Femtoampere entscheiden (B&R Industrie-Elektronik customer magazine), in: Automotion, No. 11/2015.

[12]  Apfeld, R., Huelke, M., Lüken, K. and Schaefer, M.: Manipulation von Schutzeinrichtungen an Maschinen, published by Hauptverband der gewerblichen Berufsgenossenschaften (HVBG), Sankt Augustin 2006.

[13]  Schreier, J.: Über-Automatisierung reduziert Flexibilität in der Fertigung, in: MaschinenMarkt (MM)- Das Industrieportal, available online at: http://www.maschinenmarkt.vogel.de/themenkanaele/automatisierung/fertigungsautomatisierung_prozessautomatisierung/articles/453273, last accessed: 22.07.2014.

[14]  Schmertosch, T.: Funktionale Sicherheitstechnik für Maschinen und Anlagen, in: VEMAS Industriearbeitskreis AUTOMATION, Forum INTEC, Leipzig 2015.

[15]  Federal Ministry of Education and Research: Service robots instead of nursing homes (press release 042/2016), available online at: https://www.bmbf.de/de/service-roboter-statt-pflegeheim-2727.html, last accessed: 15.05.2017.

[16]  Fachforum Autonome Systeme im Hightech-Forum (ed.): Autonome Systeme — Chancen und Risiken für Wirtschaft, Wissenschaft und Gesellschaft. Final report — short version, Berlin 2017, available online at: http://www.hightech-forum.de/fileadmin/PDF/autonome_systeme_abschlussbericht_kurzversion.pdf, last accessed: 30.07.2018.

[17]  Berufsgenossenschaft Holz und Metall (ed.): Fraunhofer IFF Magdeburg forscht im Auftrag der BGHM zur Mensch-Roboter-Kollaboration in Industrie 4.0, Magdeburg 2016, available online at: https://www.bghm.de/bghm/presseservice/pressemeldungen/detailseite/sichere-zusammenarbeit-von-mensch-und-roboter, last accessed: 30.07.2018.

[18]  Audi (ed.): Colleague 4.0, in: Dialoge. Smart Factory, 2017 edition, Ingolstadt 2017, available online at: https://www.audi-mediacenter.com/de/publikationen/magazine/dialoge-smart-factory-2017-364, last accessed: 15.05.2017.

[19]  Schenk, M. and Elkmann, N.: Sichere Mensch-Roboter-Interaktion, Anforderungen, Voraussetzungen, Szenarien und Lösungsansätze. Demographischer Wandel — Herausforderung für die Arbeits- und Betriebsorganisation der Zukunft (Schriftenreihe der Hochschulgruppe für Arbeits- und Betriebsorganisation e. V., Bd. 1), Chemnitz 2016.

[20]  Bernecker + Rainer Industrie-Elektronik GmbH: OPC UA TSN — Über die Edge in die Cloud, in: Automotion, No. 09/2017.

[21]  Hofmann, J.: Die digitale Fabrik — Auf dem Weg zur digitalen Produktion, published by the German Institute for Standardization (DIN), Berlin 2017.

[22]  Hofmann, J. (co-author): Data Leader Guide 2016, published by Connected Industry e. V., Berlin 2017.

[23]  Microsoft Corporation: Modern Data Warehouse Solutions, available online at: https://www.microsoft.com/de-de/cloud-platform/advanced-analytics, last accessed: 30.03.2017.

[24]  Fraunhofer IAIS: Big Data Analytics, available online at: https://www.iais.fraunhofer.de/de/geschaeftsfelder/big-data-analytics/uebersicht.html, last accessed: 30.03.2017.

[25]  B&R Industrie-Elektronik GmbH: Aprol PDA. B&R Industrial Electronics Company Brochure, Eggelsberg 2015.

[26]  Wildemann, H.: Modularization in Organization, Products, Production and Service (TCW-report No. 66), Munich 2014.

[27]  Plattform Industrie 4.0, May 2023. Progress report 2023: Industry 4.0: On the way to an intelligently networked industry. (Federal Ministry for Economic Affairs and Climate (BMWK), ed.) Retrieved August 20, 2023 from https://www.plattform-i40.de/IP/Redaktion/DE/Downloads/Publication/Manufacturing-X.pdf?blob=publicationFile&v=2.

[28]  VDMA e.V. April 11, 2023. Discussion paper — Interoperability with the asset administration shell, OPC UA and AutomationML: Target image and recommendations for action for industrial interoperability. Retrieved July 10, 2023 from https://www.vdma.org/viewer/-/v2article/render/78243357.

# Chapter 2

# Design of Modular Machines and Systems

The analysis of the requirements has led us to the realization that modular concepts can be the key to success in many ways. But before we take a closer look at this topic, let's take a look outside the proverbial box.

Generally speaking, we encounter modular elements of all kinds every day. We live in prefabricated houses, drive in platform-built cars, use ready-made computers, learn, work in hierarchies, and are generally part of the social system. Even our biological diversity could be seen as a large modular system in which every species fulfills a specialized function, just like every organ and every cell. Admittedly, the analogy to nature is a little tenuous because in order to really speak of modules in the sense of the definition, the property of functional interchangeability is required, which is not found in nature. Although the transition is quite fluid, as organs are already being transplanted and artificial joints produced and implanted using 3D printers.

Nevertheless, it is worth looking in this direction because we can learn a lot about structures and the functioning of modular systems from nature. Have you ever thought about why the human organism continues to function even when asleep or under anesthesia? It is due to the ability of the liver, kidneys, and many other organs necessary for viability to work completely independently. Each organ fulfills its genetically defined function by analyzing the substance to be processed — e.g. the blood with all its constituents — and carrying out the resulting activities. The organ receives and sends additional information through enzymes, among other things, creating a kind of communication system. It is a way of working that has proven itself since the beginning of evolution and that we want to keep in mind as we proceed.

We are familiar with something similarly evolutionary — albeit "only" the rapid technical development of the past 100 years — from the automotive industry. Anyone who has ever put together the configuration for a new dream car will have noticed how many versions of a vehicle can be produced in series. Starting from a basic variant, thousands of derivatives can be created using a wide range of equipment options, meaning that we have been experiencing a batch size of 1 in series production for a very long time. In order to achieve this, it was necessary to precisely structure the product itself — the vehicle — as well as all the processes surrounding production. The development of *modularization* therefore extends from the products and services, through production, to the global modularization of the production system [1]. It is therefore worth keeping an eye on this highly productive industry, as it not only sets an example for the development of modularization strategies but is also a driver of technology in many different ways.

The topic of *modularization* is by no means new in mechanical and plant engineering. Mechanical components such as screws, gears, motors, and many others are nothing more than modular components that are selected and assembled from standardized catalogs depending on the functional task. In electrical engineering, for example, cables, switches, and fuses can also be described as modular components. But why do we still want to devote so much attention to this topic?

The reasons for this can be found in almost every paragraph of the first chapter. Regardless of whether individualization of products, machines, and systems, their increasing complexity, intelligent safety concepts, or increasing efficiency — the methods of modularization methods contribute to the solution in every single aspect. Added to this are the increasing requirements arising from the digitalization of production, with the result that many purely mechanical components are now becoming mechatronic, thus fueling the trend toward the intelligent structuring of many production processes. And because automation technology plays a key role in this environment, the topic of *modularization* will also determine the other topics in this book.

## 2.1 Definition and properties of modules

In [1], the *modularization is described* as a method *in which complex overall structures are subdivided into individual, separately coordinatable modules in order to subsequently reassemble them into a complexity-reduced overall structure. The individual modules consist of a large number of elements that are compatible with each other by means of a suitable hierarchy.*

This very apt definition contains a number of terms that will play a crucial role in this chapter. Before we therefore shift our attention to the design and project planning of modular systems, the essential principles and approaches of modularization should be explained.

### 2.1.1 Modularity

*Modularization means breaking down an overall system into meaningful components known as modules or building blocks. Their form and function are designed in such a way that they can be easily joined together and/or interact via appropriate interfaces.*

This definition can be found in one form or another in numerous publications, such as [1–4] or [5]. Since modularization, as described in the introduction, applies to almost all areas of production processes, we therefore solely focus on the essential properties of *mechatronic components* at this point.

Under this premise, a *module can* be generally defined as follows:

- A module is a functional element of an overall system.
- It has material, energy, and communication interfaces.
- It can be replaced with a functionally identical module for the rest of the system without retroactive effect.

In addition, [5] defines various types of modules in plant engineering whose basic approach also appears relevant to mechanical engineering. These are in detail as follows:

- **Type 1 Autonomous modules**
  Self-contained units with their own automation, which function independently and autonomously and are connected via a standardized

interface to a central control system for operational data acquisition and monitoring.

- **Type 2 Integrable modules**
  Units that are firmly defined in terms of function and area of application, which are integrated into the overall system in terms of materials, energy, and automation and are directly influenced via a central control system. This enables module-spanning and comprehensive control from the perspective of the overall system.

- **Type 3 Modular modules**
  This means modularization on several levels. Autonomous and integrable modules can be further modularized in their design, and thus their functional structure can be changed without affecting the overall system. The integration of a module basically corresponds to that of integrable modules.

The self-sufficiency of autonomous modules is the main difference between autonomous and integrable modules. Integrable modules, on the other hand, require the surrounding overall system in order to function at all. If we consider a complete machine as an autonomous module, then this description applies in full because the machine can work completely autonomously on its own and the production control system has no direct controlling influence on the machine. However, it can query data or send orders via services, whereupon the machine changes its operating status or executes technological functions. Internally, the machine can also be understood as a modular module if it has a modular structure, whereby the individual modules can be integrated, and there is a central control system that controls the functions and internal operational processes.

In addition, autonomous modules can also be described as a *black box*, where only the external behavior, i.e. the functional input and output, is relevant. This results in a very important module characteristic — the *encapsulation*.

It can also be deduced from the definition of type 3 that there are modules that represent a subset of functions of a more complex module. The functions of the sub-modules can and should differ if the overall function is not affected.

Let's look at a *perfect binder*[1] as an example machine. Among other things, this contains the mechatronic component *glue station*[2] for the application of adhesive, with which an adhesive is applied to the pre-milled book spine either as a hotmelt, PUR, or dispersion adhesive, depending on the technological requirements. While the external *adhesive application* function of this module is fixed, the internal structure only differs in the area of adhesive dosing and application due to the modular design of the gluing station. And because all external interfaces of the glue station are identical regardless of the type of adhesive, the module there-fore behaves like a *black box*.

We can see from this example that type 3 variants can have differenti-ated sub-functions with the same external behavior. For this reason and in order to make a better distinction, we want to use the terms *base module* and *derivative*:

- **Base module**
  Modular part of a complex system with firmly defined external func-tions and interfaces but variable sub-functions.
- **Derivative of a module**
  Variant of a base module formed by embedding one or more variable sub-functions.

In our example, the autonomous module *glue station* is the basic mod-ule, which is designed in various derivatives with correspondingly inte-grable modules for different types of adhesives.

Now we need to address the question of the level of complexity at which we speak of modular modules and of modules at all. When is a module really a module and no longer a component?

A look at the theory of process analysis helps us to answer this ques-tion. There is the term *"balance boundary"*, and this refers to a method of delimitation for analyzing and evaluating a system or subsystem, e.g. for a material or value stream analysis. A complex system is subdivided into

---

[1]Perfect binder is the industry-standard term for a machine used to produce a book from an assembled book block and a cover [31, 32].

[2]The correct technical term is *adhesive application system*, for which the term *glue station* is used as a synonym in the following.

many smaller balance areas, whose inputs and outputs are analyzed, and the results analyzed and evaluated using different methods. The size of the balance areas and the course of the associated balance boundaries are defined according to various criteria.

If we apply this procedure to a production plant, a balance sheet boundary can be a production hall, a single machine, but also a finished product. In the production hall, the individual machines can be viewed as modular components that each make their contribution to the manufacture of a product within an overall system. If we look at the product, for example, the energy balance can be represented by the inputs and outputs of all energy flows of each production step. At the beginning of modularization, the balance sheet boundaries must be sensibly defined, which results in balance sheet spaces that we want to call modules in the sense of mechatronic units. It depends on many factors and the methods where these boundaries lie and how large the balance areas will be. We further look at this in Section 2.3.

## Module

A *module* is within the defined balance sheet boundary *machine* or *plant* a more or less independently operating *encapsulated* component with the desired functionality and defined interfaces. We distinguish between *autonomous* and *integrable* modules, which in turn can take the form of various *derivatives*. Autonomous modules can be regarded as a *black box* within an overall system.

In order for a modular system to work and be treated as such, two important properties of modules are required, which we deal with in the following section.

### 2.1.2  Functionality

*The function of an object describes the specific task for which it exists. The totality of all functions is referred to as functionality.*

This clearly defines that each object — in our case each module — is characterized by its *functionality*. A hard disk stores data, a webcam provides a video stream, and the gluing station applies glue to the spine of a book in a perfect binder. However, with *f* more or less complex modules in mind, we need to take a closer look at the term *functionality*.

Let's take the example of the hard disk in a computer: From the outside, it is a simple mass storage device with more or less storage space. However, this description also applies to a USB stick or an SD card. Apart from their design, all three and many other models in the mass storage class have the same external appearance. The glue station also has a very specific task to fulfill at its fixed location in the machine. The way it operates internally should be indifferent from an external perspective for us to view it as a black box. For classification in the overall system, we are therefore only interested in the *external functionality*, which we also refer to as the *basic functionality*.

However, it is obvious that each *derivative* of a *base module* implements the basic functions in different ways depending on its actual design. Contrary to the SD card, the hard disk has a rotating disk that is driven, a read/write head that has to be positioned, and many other components that must perform the corresponding tasks. In the following, we use the term *internal functions*.

### ✓ Functionality

Each *module* has a *basic functionality* that enables it to fulfill the tasks assigned to it in an overall system. To do this, it uses *internal functions*, which can vary in type and scope depending on the *derivative of* a *base module*.

## 2.1.3  State and state changes

*The state of an object is understood to be the complete description of its current properties in relation to its use.*

These or similar definitions are used in all scientific disciplines. Therefore, the *relation to its use* is embedded in the definition because all properties and information of an object or system are rarely of any interest. If, for example, the aggregate state of water is to be described, then the structure of the hydrogen atom is not important, but temperature and viscosity are.

There is another important variable in this definition — the *temporal reference*. This highlights that the state of a dynamic system is subject to constant change (or not, but this is also a statement about dynamics). These changes are referred to as *state transitions*. In this context, computer science uses the term *state space* to describe and represent the set of all states and their transitions.

Consequently, in the automation of technical systems, the control design is represented in the so-called *state machines*.[3] The information for a state is displayed at the current time. A *state transition* causes a change in the state of the automaton and occurs when the logical conditions defined for this transition are fulfilled. The *state machine* recognizes four types of actions:

- **Incoming action**
  Output occurs when entering a state
- **Outgoing action**
  Output takes place when leaving a state
- **Input action**
  Output depends on the current status and the input
- **Transitional action**
  Output is dependent on a state transition

A typical representative of *finite state machines* is the variant of *transductors* and the *Moore automaton*. An automaton is said to be *finite* if the set of acceptable states is finite. In *transductors*, inputs are recognized depending on a state and outputs are generated in the form of actions. The *Moore automaton* only uses incoming actions, while the outgoing actions depend on the state itself and the input. This makes it easy to understand and is therefore frequently used.

The simplified control of a belt switch with a Moore automaton will be designed to illustrate this. Table 2.1 contains the defined states with their properties, in- and outgoing actions, as well as the switching conditions and subsequent states. Figure 2.1 demonstrates the model of the automaton with its transitions.

---

[3]Similar terms with approximately the same meaning are the *finite automaton* or the *state machine*.

**Table 2.1.**    Overview of the states and actions of a simplified switch control system.

| | Properties | | | | | | | | | Actions | | Transition condition | Subsequent state |
| | Inputs | | | | Drive | Outputs | | | | | | | |
| Condition | ES Li | ES Re | Start | Quit | Drive | Yellow | Blue | Green | Red | Entrance | Output | | |
|---|---|---|---|---|---|---|---|---|---|---|---|---|---|
| Ready | x | x | 0 | x | Active | Blink | 0 | 0 | 0 | Yellow = blink | Yellow = 0 | Start = 1 | Right |
| Right | 0 | 1 | x | x | Active | 0 | 1 | 0 | 0 | Blue = 1 | Blue = 0 | Change | Drive left |
| Left | 1 | 0 | x | x | Active | 0 | 0 | 1 | 0 | Green = 1 | Green = 0 | Change | Drive right |
| Drive left | 0 | x | x | x | Left | 1 | 1 | 0 | 0 | Blue = 1 Yellow = 1 | Blue = 0 Yellow = 0 | ES Li = 1 | Left |
| | | | | | | | | | | | | Error | Malfunction |
| Drive right | x | 0 | x | x | Right | 1 | 0 | 1 | 0 | Green = 1 Yellow = 1 | Green = 0 Yellow = 0 | ES Li = 1 | Left |
| | | | | | | | | | | | | Error | Malfunction |
| Malfunction | x | x | x | 0 | Error | 0 | 0 | 0 | Blink | Red = blink | Red = 0 | Quit = 1 | Ready |

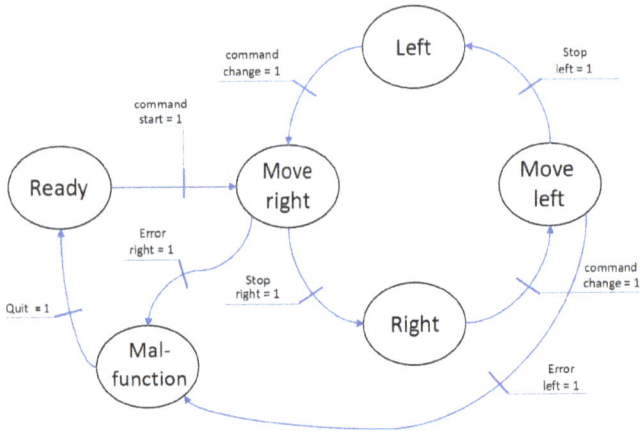

**Figure 2.1.**    Moore automaton of a simplified switch control system.

It is noticeable that no input is provided for the *change* command. So where does the request to change direction come from? Section 2.2.3 offers a possible solution.

> **Condition**
>
> A *state* represents all the currently relevant properties of a system at a specific point in time. *Changes of state* bring about the transition from one state into a subsequent state. The representation takes place in *state machines*.

The term *state* is explained in detail in Section 2.2.

### 2.1.4 Compatibility

*In technology, compatibility refers to the interchangeability of assemblies under the condition of equivalent properties.*

This definition is considered common knowledge and is included in numerous dictionaries. Reference [6] defines *compatibility* as general compatibility or tolerance. One example is data processing, *which requires devices, data carriers, data, and programs to be interchangeable or linkable without special measures.*

If we imagine that the individual components of a computer are nothing more than autonomous or integrable modules in a wide variety of derivatives (a 120 GB hard disk behaves in the same way as a 250 GB hard or a silicon disk as mass storage), then we quickly realize what compatibility means: connect — switch on — go without any further action or complex adjustments.

Let us take a look at the example of a *USB interface*.[4] All devices that support this communication standard work according to the *plug and play* principle, which is a pleasant procedure due to its functionality, at least in most cases. That is *compatibility*.

In contrast, we register an *incompatibility* immediately when something is connected — or is supposed to be connected — and does not fit or does not work. The USB3 plug does not fit mechanically into the mini-USB2 socket, or a program does not run under the Linux operating system because it was developed for Windows. However, with an appropriate adapter cable, the USB3 device will work on an older USB2 host but with a lower data rate. The USB3 protocol is therefore *downwards compatible*. In contrast, there is *upward compatibility* for USB2 devices on USB3 hosts, which means that if a PC only has a USB3 port, a USB2 hard disk can be operated on it.

This example illustrates that a module is only truly compatible if it completely fulfills the specification defined for the intended use. This also includes consideration of the functionality. After all, a module or a derivative of a module is only truly compatible if it has the basic functionality required for the intended use, i.e. *functionally compatible*. For example, a module can have all the necessary physical interfaces and therefore be compatible with a system. However, if it does not fulfill its intended task or if the system cannot do anything with this component, then it is still *incompatible*.

We recognize from this that 100% compatibility is not that simple. For the creation of modules, it is therefore essential to precisely define the states and expected functionality at the balance limits. It is crucial to strictly adhere to the development process and document this procedure.

---

[4]Detailed information can be found at http://www.usb.org.

☑ **Compatibility**

A module is *compatible* if it can be replaced with an equivalent module without any special measures. It is *downward compatible* if it is compatible with older versions and their interfaces, and it is *upward compatible* in the same way if it is also functional in a newer environment. In order for a module to be integrated into an overall system, it must be *connection-compatible* and *functionally compatible*.

In practice, the issue of *compatibility* is simplified through a large number of standards, specifications, guidelines, recommendations, and even laws in the interests of the respective users. The reality is that technical progress, especially in information and automation technology, has increasingly overtaken every standard by the time it is published. In addition, the requirements of the various industries differ, and everything is additionally overlaid by more or less comprehensible company interests, such as know-how protection and competition. The result is that compatibility in the industry cannot function in the same way as, for example, a USB device. Or does it?

The following sections shows us that the integration of mechanics, electronics, and software into mechatronic units, on the one hand, and the advent of digitalization in production, on the other hand, require us to address the issues of *modularity, functionality,* and *compatibility* in a much broader sense.

## 2.2 Modularity in the context of Industry 4.0

So far, we know that the introduction of IT technology into production halls leads to machines and systems also acting as autonomous modules and communicating directly with each other. And because we also know that this works quite well with USB devices, it is obvious that this functional principle is also of interest to industry. *The "USB principle for the factory should enable the integration of a new machine into a production process as easily as the connection of a printer to a computer nowadays"* [7].

It's not that far now. The online library of the German *Industry 4.0 platform*[5] and the VDMA[6] contain numerous publications that describe exactly how to get there. The following documents, which also serve as a basis for further considerations, should be referred to as examples:

- *Industry 4.0 — Technical Assets Basic terms, concepts, lifecycles and management* [8] based on the properties of assets and their classification in the digital production of reference models and architectures.
- *Reference Architecture Model Industry 4.0 (RAMI 4.0) An introduction* [9]
- Introductory description of RAMI 4.0.
- *Structure of the administration shell, further development of the reference model for the Industry 4.0 component* [10]
- Description of the further development of the content of the RAMI 4.0 architecture and definition of the administration shell of an I4.0 component.
- *Further development of the interaction model for Industry 4.0 components* [11]
- Treatment of the interaction capability of I4.0 components as an essential basis for the functioning of digital production.
- *Industry 4.0 Plug-and-Produce for Adaptable Factories: Example Use Case Definition, Models and Implementation* [12]
- Presentation of the current state of research in application and implementation examples.
- *Discussion paper — Interoperability with the asset administration shell, OPC UA and AutomationML: Target image and recommendations for action for industrial interoperability.*
- Publication of a joint target image and recommendations for action for industrial interoperability by the associations AutomationML e.V., IDTA, OPC Foundation, and VDMA.

To gain more clarity in this context, it is necessary to represent and treat modules in the form of mechatronic units as objects, just like products in general as objects. Thus, we look at the vocabulary of *object-oriented programming (OOP)*. This is useful insofar as the modularization approach is genetically located in OOP and the description methods are

---

[5] Official homepage, http://www.plattform-i40.de.
[6] Official homepage, https://www.vdma.org.

based on the same motivation, namely, to reduce and control the complexity of software projects. However, it is important to bear in mind that real objects differ in some aspects from pure software objects. We want to work out these divergences by making use of some OOP concepts and applying them to real systems. In doing so, we consider some — and only those relevant for further consideration — basic terms that are described in the same way in most sources. The definitions used and partially quoted here are taken e.g. from [13, 14], where they can be read in more detail.

With this knowledge, we dive into the terminology for the definition and interaction of I4.0 components and use practical examples to show how the implementation of object-oriented modularization can work.

### 2.2.1  Objects and entities

In [13, 14], an *object* is described as follows:

*In general, an object can be regarded as an entity that is clearly separated from its environment and has its own identity. It reacts to stimuli from its environment (e.g. changing its own state) or sends signals to the environment itself.*

*An object consists of data and has access to methods, for example to change its state.*

*The consistent implementation of the object-oriented view only allows data to be changed using object-specific methods.*

*Hiding the data and methods to a unit is called encapsulation.*

*A class is a construction specification for objects and defines common properties, functionalities (methods), semantics and relationships to other objects or classes.*

This definition also applies in full to modules in the form of *mechatronic units* because

- the systems are delimited,
- react to input signals,
- process them according to their internal functionality,
- then change their condition,
- can only influence their condition through their own functions, and
- have external functionality.

If mechatronic modules are only visible to their environment via their *external functions* as *functional modules*, they also meet the definition of *encapsulation.* In addition, *basic modules* and their *derivatives* can be summarized and described in a *class* and a derivative can be regarded as an *instance* from this perspective.

Furthermore, *functional modules* from the perspective of data modeling can be described as *entities*, which are defined in [13] as follows:

- *An entity is a clearly identifiable object in the real world.*
- *An entity has many properties, of which only those are visible that are of interest for a specific section of reality.*
- *A property describes an essential characteristic of an entity and is assigned a property value that originates from a predefined value range.*
- *The abstract description of a set of entities with the same properties is called an entity type or class.*
- *A primary key is defined in the set of properties, the value of which is always unique and with which an entity can be uniquely identified over its entire lifetime.*
- *Relationships are used to link two or more entities with each other.*
- *The cardinality of a binary relationship indicates how many entities of one entity set are related to how many entities of the other entity set.*
- *The individual relationships of the entities are visualized in a relational model, the entity-relationship model (ERD).*

There are numerous forms of representation for an *ERD* in the literature. In the following, we want to represent entity types as a rectangle and give them a name. Their properties marked with names are added and connected to the entity type by lines in a circle or an ellipse. The primary key is underlined in a separate block. The relationships between the entities are represented by lines labeled with the relationship type and cardinality. Figure 2.2 shows an ERD section, which displays the entities *vehicle* and *customer* with assigned properties. Both entities are uniquely identified via the primary key *serial* or *customer number.* The *cardinality* specification indicates that each vehicle can only be assigned to one customer, but one customer can be assigned to none or several vehicles.

Remarkable about this definition is that it focuses on one *section of reality* because it implies that there are other entities with properties that are not relevant at the time of consideration. For example, the *drive* property itself describes a network of relationships between various entities of the drivetrain such as the engine, gearbox, and differential, but this is not

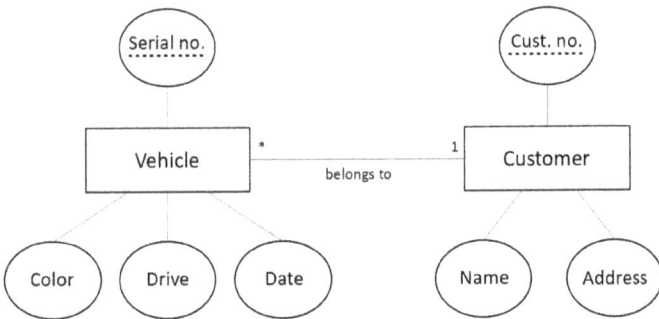

**Figure 2.2.** The illustration shows an assignment of property types to an entity type and a relationship to another entity.

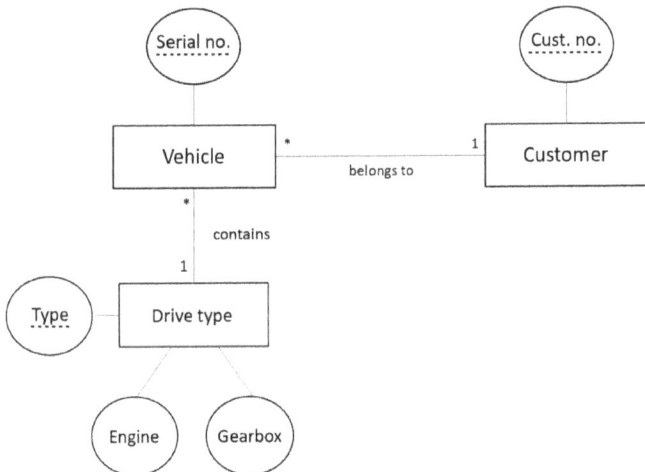

**Figure 2.3.** This extended slice of reality shows relationships as they may be relevant to the service.

visible in the section of reality shown. The model in Figure 2.2 could therefore also be called *sales* and the ERD model in Figure 2.3 is called *service*. It can also be seen that only those properties are shown that are relevant to the respective reality section.[7] In this case, the service department is only interested in the individual customer vehicle that has a certain type of drive.

---

[7] Cf. Section 2.1.1.

The example of the *vehicle* entity makes it clear that the definition of the entity can easily be transferred to a functional module in the form of a mechatronic unit. It is a real existing object, has constructive and functional properties, and is related to other modules. The *vehicle* can also be regarded as a product in the production process or as a mechatronic unit with corresponding functions. This shows that different sections of reality can arise for one and the same entity, depending on the perspective from which it is viewed. In the following, we use this approach to illustrate the interaction of processing units, a machine of a plant, and/or its modular components with each other and with a product to be processed.

To illustrate this, we again use the *perfect binder* with the *glue station* module, and the product is the *individual photo book*. Figure 2.4 shows these components as entities with some selected properties and relationships. For example, a perfect binder that contains at least one gluing station can process a certain number of photo book types, and a gluing station can be assigned to only one perfect binder or be on standby outside of it. If there are several types of gluing stations, an extended ERD would have to show that a certain type of photo book can only be processed in a perfect binder if a certain type of gluing station is also available.

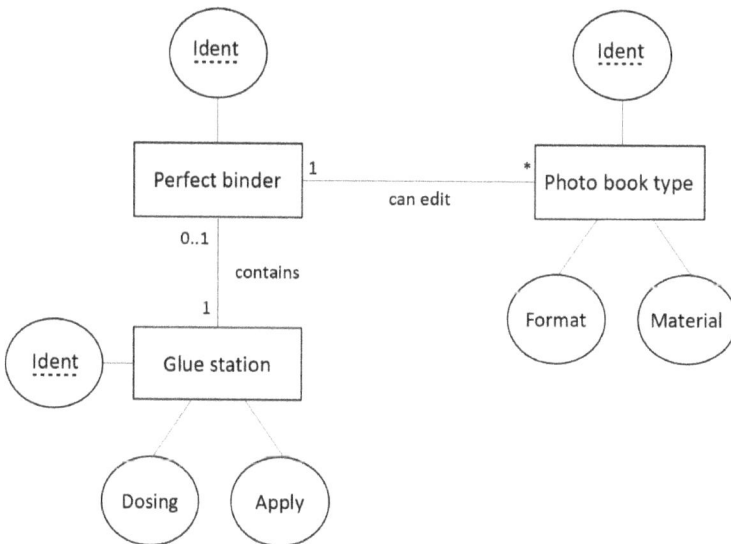

**Figure 2.4.** Example of an ERD for the production of a photo book in a perfect binder and a glue station as a functional module of the machine.

## 2.2.2 Methods and functions

In Section 2.2.1, it was stated that an object can be changed by accessing *methods* to change its state. A method is defined in more detail in [13] as follows:

*A method is an algorithm that is assigned to an object and can be processed by this object.*

If we project this view onto a mechatronic system, we only need to understand the methods as the external functions of a mechatronic component and assign them to an entity. If we use this assumption to describe the functional module *glue station*, it has properties in the form of *basic functions*. The derivatives of this module can also be summarized in a class *glue station*, as they have the same externally visible functions, namely, the *basic functions* (Figure 2.5).

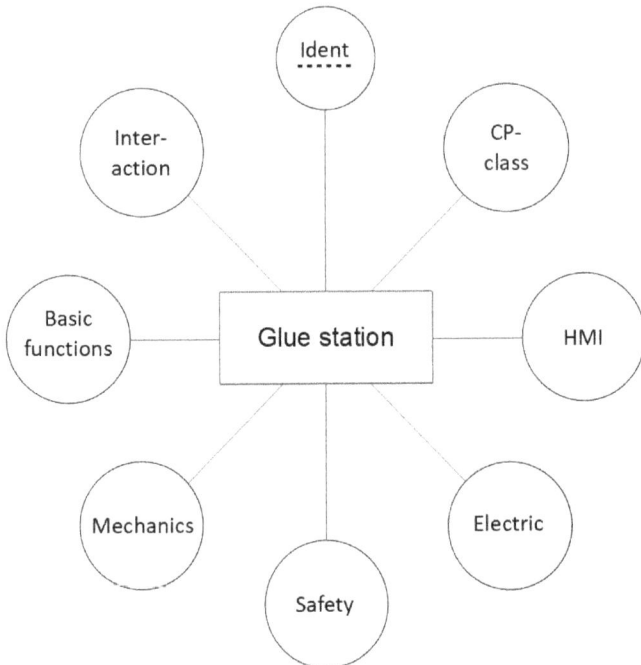

**Figure 2.5.** Representation of the mechatronic component glue station as an entity with selected design properties and the external functions relevant for operation in a perfect binder as an additional property.

In this ERD, in addition to the basic functions, the *glue station* is also assigned the mechanical and electrical properties as well as the operating functions for *human–machine communication* (HMI) that are required for its integration into the overall perfect binder system. It is irrelevant whether the gluing station has its own HMI device or whether these functions are displayed on another (central) operating system.

### 2.2.3 Messages and services

In [13], *communication* between objects is defined as follows:

*Objects communicate with each other via messages.*

If a module receives a *message*, the object responds with a clearly defined method to change its state because the receipt of a message is a request to the object to trigger a suitable method to *change its state* [14]. A message can be a complex data telegram, the value change of a variable, or, in mechatronics, the status change of a sensor signal.

In the context of Industry 4.0, the functions of the system components are accessed via *services*. As a result, a functional module, if it should act as an I4.0 component, must be a *service participant* of a *service platform* [8], and a *service* is described as a *delimited function that is offered by an entity or organization via interfaces.*

Reference [15] considers the process of merging mechanical and plant engineering with IT and automation in more detail. Accordingly, we must distinguish between data traffic in operations (the technologically required control and regulation tasks) and data exchange for technical organizational measures such as data analysis or service tasks. Therefore, the context in which services should be requested or data exchanged must be considered. In [15], a description structure is presented on the basis of which a service-oriented interaction between different entities can be organized and forms the basis for all I4.0 interaction models (Figure 2.6). It describes a service with an *identifier*, a *service type identifier,* and the *service parameters*. This identifier is used to call the service to process the *service input data* sent with the call. Finally, the service must return the results via *service output data*.

From the perspective of a technical device, the services represent nothing more than the calls to its external functions. These are primarily the basic functions defined in the class definition. For each derivative of a device belonging to a class, there are further functions that can be called

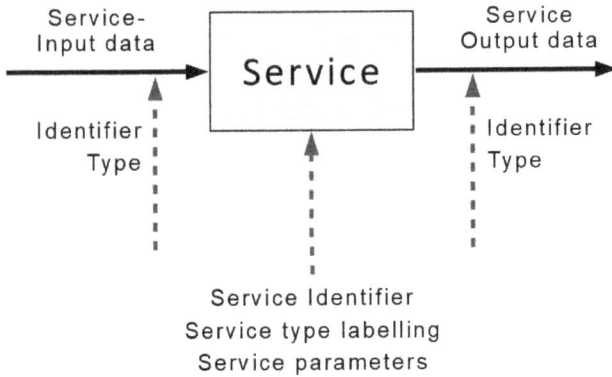

**Figure 2.6.**    Example of a description structure of services.

*Source*: According to [15].

via additional services. For example, in the case of the *glue station*, this is the glue type. The information about the range of services for I4.0 components is stored in the so-called *manifest*,[8] which is called up when an object is plugged in. The manifest is located within the *administration shell*,[9] which is used *to access the functions of the system components via well-defined services* [8].[10]

Let's look at the example of the switch control from Section 2.1.3. If the track switch is managed as an I4.0 component, the command to change direction could be defined as a service call. The input data then contains the desired direction, and the output data contains the success or, in case of an error, the failure as information.

Especially in the context of the interaction of I4.0 components is still under *service users* and *service providers*. This difference can best be described using the production chain for manufacturing the individual photo book.

A control system is responsible for the organization of production, which acts as a service user in relation to the production process and views the individual machines as service providers. From the perspective of the product, the product itself is a service user and the machine is a

---

[8]There is currently some disagreement within the I4.0 committees as to whether a *manifesto* is fundamentally necessary. After all, this functionality is already largely covered by the OPC UA protocol.

[9]Cf. I4.0 components, "The Administration Shell" (Section 2.2.4.4).

[10]Opinions also differ somewhat on this point, as service-oriented communication is already taking place through the use of OPC UA alone.

service provider, as the product undergoes a change in its condition through its processing condition. This becomes easier to understand if you imagine that when a product arrives at a production station, it transmits information about its status to the machine via a QR code or RFID chip and thus requests a service for further processing. For a modular machine, on the other hand, the specific functional module is the module is the service provider, as it performs a specific service within this production unit in the form of its functionality.

## 2.2.4 The I4.0 component

For an entity in the Industry 4.0 environment, the term *I4.0 component* has become established. The conceptual framework for the description of an I4.0 component is the *Reference Architecture Model Industry 4.0 (RAMI 4.0)*. This three-dimensional layer model connects the key elements in the context of Industry 4.0 so that I4.0 technology can be systematically categorized and further developed [16]. Even if, as of November 2023, there is still a need for further research before the first industrial applications can be realized [17], it still seems appropriate to use the current structural models for the description and interaction of I4.0 components in modularization concepts to be created today.

In the following sections, in addition to the basic terms of OOP already introduced, we take up the terminology with which I4.0 components are described in more detail. The following documents form the basis for this:

- *Status Report Industry 4.0 - Technical Assets — Basic terms, concepts, lifecycles and management* [18]
- *Status report — Further development of the interaction model for Industry 4.0 components* [11]
- *Working Paper Industry 4.0 Plug-and-Produce for Adaptable Factories: Example Use Case Definition, Models, and Implementation* [12]
- *Discussion paper — Interoperability with the asset administration shell, OPC UA and AutomationML: Target image and recommendations for action for industrial interoperability* [38].

We want to underpin the emerging overview of I4.0 terminology with suitable examples for better understanding. To do this, we once again use the perfect binder in the context of digital production for the manufacture of individual photo books.

### 2.2.4.1 *The reference architecture model Industry 4.0 (RAMI 4.0)*

To ensure that the entire value chain can be represented and managed in a sustainable and consensual manner in the digital production environment, it is necessary to structure it and map it in standardized reference models. In the context of *Industry 4.0, reference* architectures are also required in addition to the reference models, which form a conceptual framework for system design. The reference models describe the construction principles belonging to this framework. The reference architecture relevant to I4.0 systems is defined in the *RAMI 4.0 model* (Figure 2.7) and published as DIN SPEC 91345 [19]. In this architecture model, *technical assets* are the basis of this architecture model and are therefore examined in more detail in the following sections.

To understand this model, it is important that it represents technical assets by category over the entire life cycle and their roles. In [16], the structure of RAMI 4.0 is explained as follows:

- **Axis "Hierarchy Levels"**
  These hierarchy levels, standardized in IEC 62264, represent the different functionalities in a factory. The product to be manufactured

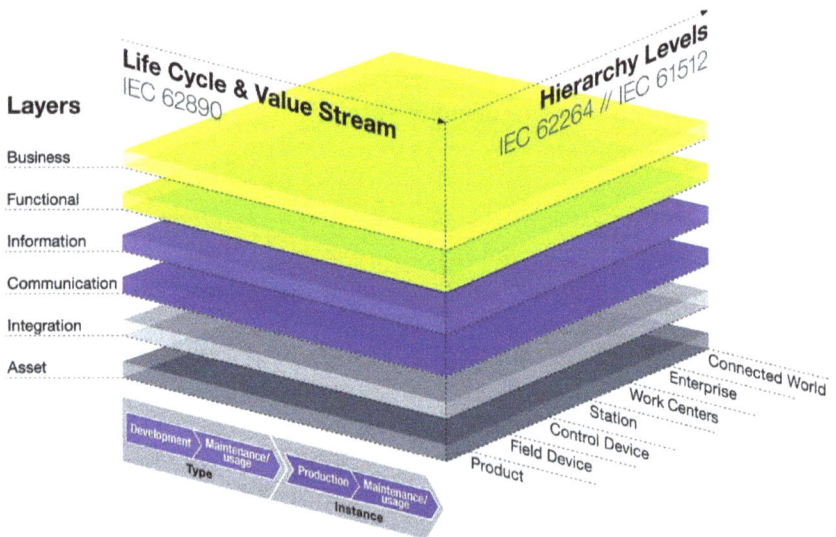

**Figure 2.7.**   The reference architecture of RAMI 4.0.

*Source*: According to [19].

represents the lowest hierarchy level and the Internet of Things and services as the *Connected World* represents the highest hierarchy level.

- **Axis "Life Cycle & Value Stream"**
  This diagram represents the life cycle of systems and products based on IEC 62890 "Life-Cycle Management". A distinction is also made between *type* and *instance*. A type becomes an instance when development and prototype production are complete, and the actual product is manufactured in production.
- **Axis "Layers"**
  With the help of the six layers on the vertical axis of the model, the IT representation, i.e. the digital image of a product or machine, for example, is described layer by layer in a structured manner.

The RAMI 4.0 model thus unites the different user perspectives and creates a common understanding of Industry 4.0 technologies. It is therefore the basis for all further necessary developments and standardization regarding identification models, uniform semantics, uniform communication protocols, and the *quality of services* (QoS), which are important for the time synchronization, real-time capability, or reliability of Industry 4.0 components [16].

The RAMI 4.0 model in its entirety is certainly less relevant for manufacturers of device technology and software components that are to be operated in a digital production environment. However, it is important to classify the *assets* within this model and to consider the adjacent layers and axes. Device technology with IT security functions will therefore be in the *Communication* layer, whereby the interfaces to the *Integration* and *Information* layers must be taken into account. Machines and systems, on the other hand, are classified at the lowest level, just like products. This categorization is therefore one of the first design decisions when developing an I4.0 component. In the following, we therefore concentrate primarily on this level, as this is where the closest connections to the automation technology of production systems exist.

### 2.2.4.2 *Technical assets*

By definition, technical objects in the real world can be uniquely identified and are therefore an entity in the sense of OOP (see Section 2.2.1). In addition, they are manufactured for a specific purpose and have a functionality required for the production process. For example, this can be a machine, a functional module of a machine, software, or a product to be manufactured.

In principle, technical objects go through a *life cycle* and a closely linked *value progression*. Entities for which these aspects play a role are referred to as *technical assets* or simply *assets*. They are represented in the lowest level of the RAMI 4.0 model. Technical assets can be part of the *physical world* (material assets) or the *information world* (intangible assets). Material assets are, for example, machines, components, tools, computers, and their storage media. In contrast, plans, drawings, and standards are intangible assets. Class models are also assets that can be instantiated. For example, a mass-produced machine or the derivative of a functional module is an instance, as it is manufactured according to the plans stored in a class model.

Using the example of the perfect binder, the machine can be viewed entirely as an asset. If the module *glue station* is also operated at a docking station to make it ready for operation (e.g. by preheating in the hotmelt version), it may be interesting for the production process to know whether it is ready for operation or not and how long it takes to reach this state. In this case, the module itself can also be managed as an independent asset. For example, it can be in energy-saving mode as long as there is no product in the production cycle that should be manufactured using this technology. In this case, the production control system issues a service call to the module to make it ready for operation, and when this is achieved, the operator receives the service call to convert the perfect binder.[11]

Figure 2.8 schematically illustrates the life cycle of a technical asset.[12] It begins with an *order* for the manufacture of a product, e.g. a perfect binder. After the subsequent *manufacturing process* (development, design, procurement, production, quality control, etc.), the machine exists; it is basically functional but not yet ready for use. In the subsequent *provisioning phase,* the machine is delivered, assembled, and commissioned at its final destination. The machine is now ready for operation and can fulfill its functional purpose; the phase of *utilization* as a technical system in the production process begins.

---

[11]At this point, it becomes clear how important it is to classify the asset in RAMI 4.0 at the very beginning.

[12]The four phases of RAMI 4.0 are supplemented by the *order* and *disposal* phases.

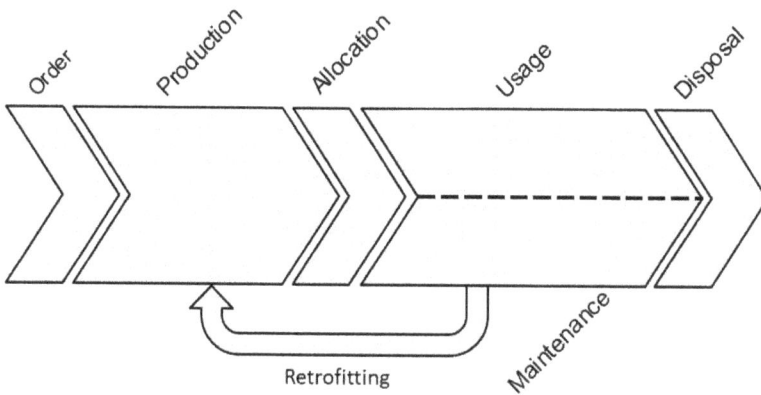

**Figure 2.8.**   Illustration of the life cycle of technical assets.

*Source*: According to [18].

From this point on, the machine is also a case for *maintenance*, from whose point of view it continues to represent a product that must be maintained in its functionality. It may be the case that the machine is fundamentally overhauled or modernized in a general overhaul. During this *refurbishment phase,* activities are carried out on site, by or at a service provider or the manufacturer, which can also lead to a return to the status of manufacture. Therefore, the technical device may be in such a loop several times during its life cycle. The machine assumes the status of *disposal* only when it is completely removed from the production process which ends with its final dismantling and scrapping.

This life cycle also includes the *value progression* of an asset, which rises steadily during the production phase and falls in the utilization phase due to aging effects until it finally ends at zero again after disposal. Maintenance and refurbishment can increase the value again during the utilization phase (Figure 2.9).

The representation of the assets is displayed in the RAMI 4.0 architecture model on the lowest horizontal level. On the axes of this level, the assets are displayed according to *categories* and further organized into *types* and *instances*. An asset retains this assignment over its entire life cycle (Figure 2.10).

However, the situation is different if you consider an asset according to its *role* in a technical system. Two different roles are distinguished, depending on whether the asset itself part of the technical system is or

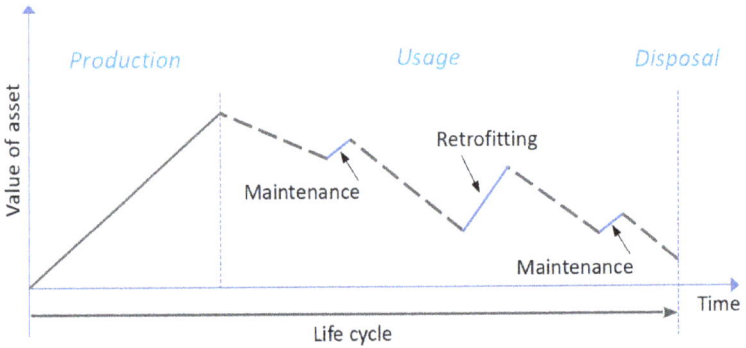

**Figure 2.9.**   Value development of technical assets over the life cycle.
*Source*: According to [18].

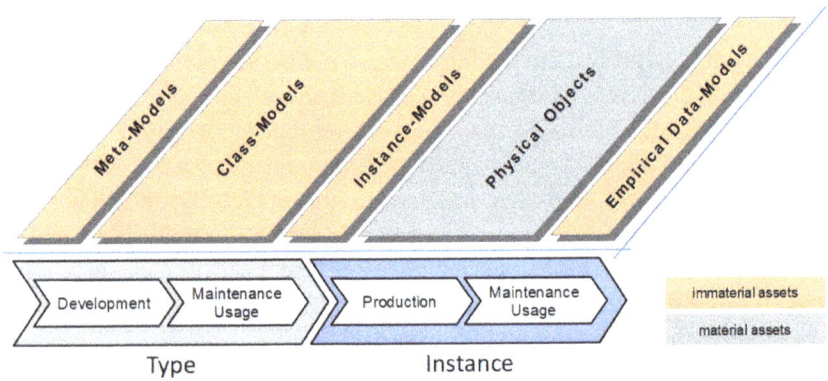

**Figure 2.10.**   Asset categories within the RAMI 4.0 model.
*Source*: According to [18].

uses it for a change of state. This consideration is important because every technical system is initially a product itself and must be treated as such.

In the RAMI 4.0 model, the assets are categorized according to their current roles (Figure 2.11). This is accomplished in two parts. The first part contains assets in their role as *user* or *product,* and the second part represents assets in their role as *furnishings of the technical system,* where they are also shown in more detail [18]. This means that an asset can be managed comprehensively as an I4.0 component in all technical systems.

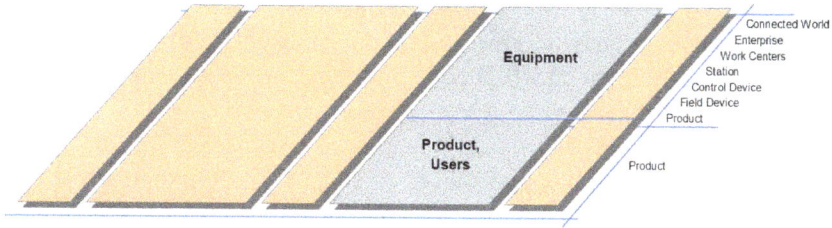

**Figure 2.11.**   Role-specific classification of assets in the RAMI 4.0 model.
*Source*: According to [18].

This view can be illustrated very well using the example of a *rolling tool* as a component of a *profile rolling machine* if both the manufacture of the machines and the production and reconditioning of the rolling tools take place in a production plant. In the manufacturing phase of the rolling tool, it has the role of a *product*, which is the same role as the associated machine. In the utilization phase, both become an *item of equipment*. If the tool is sent back to the manufacturer for reconditioning at a later date, it becomes a *product* again. However, if this takes place in the operator's own workshop, it remains an item of *equipment*. In both roles, the tool is in the *refurbishment phase*, which means that its position in the life cycle remains clearly defined. In contrast, the individual photo book functions as a product throughout its entire life cycle. In the following section, we now look at the mapping of assets in the information world.

### 2.2.4.3 *Assets in the information world*

In RAMI 4.0, the physical assets are categorized entirely at the lowest level. All levels above are assigned to the *information world*. In the following, we restrict ourselves to the next three levels, which are explained as follows in [18]:

- The *integration level* describes the different views of the assets in the base level.
- The *communication level* describes the communication relationships between the physical carriers and the information world and between the objects of the information world itself. This includes, for example, all protocols and formats.

- The *information level* comprises all aspects relating to the storage, exchange, or processing of information. An example of this would be communication via an RFID code or, in the case of a communication-capable component, via a server.

It is crucial for the management of assets in the information world that their existence and identity are unknown until they are disclosed. The content and scope, the level of awareness required for management is decided in the system design of the specific technical system. The following levels of awareness are specified for assets in the *integration level* of RAMI 4.0:

- **Unknown**
  The asset is generally unknown in the world of information.
- **Anonymously known**
  The information world only knows that an asset exists in a certain location. The system does not know any properties for this asset, but it can map the existence of one or more of these objects in a specific location.
- **Individually known**
  The asset has a unique name that is known in the information world and has properties. It has a suitable method, e.g. a QR code or an RFID chip, with which it can be identified in the physical world. It can also be identifiable through a suitable tracking strategy or by analyzing characteristic physical properties. The prerequisite is that the asset can be clearly assigned to a named object in the information world.
- **Managed entity**
  The asset has its own objects for management and use in the information world. In the so-called *administration shell,* all information and functions for tracking, the life cycle, and for operational management and control of the production process are provided. In RAMI 4.0, a fundamental distinction is made between the asset itself and its management.

This does not fix the condition that physical components of a technical system such as machines, systems, or functional modules as I4.0 components must be actively integrated into the system via communication interfaces. However, if they are to communicate actively as I4.0 components in an I4.0 environment or a digital factory, then direct communication with the technical system takes place via a separate administration shell. An asset can only function as part of the digital IT system in an I4.0-compliant manner when it is able to communicate. For our perfect binder, this means that both the perfect binder itself and the gluing station must be managed as an entity, and both have to communicate via their own administration shell.

The product, which is manufactured in a perfect binder, on the other hand, can be treated differently. If, for example, it is the individual photo book equipped with a QR code, then it is *known* as an *individual* product. When it reaches a processing station, this code can be read, and the production system can either request the required data or it is already contained in the code itself. If, on the other hand, the perfect binder produces paperbacks as serial products, the individual product may only be *known anonymously*.

The *level of awareness* and *communication skills* are described by the combined *CP-class* (Communication & Presentation) (Figure 2.12).

From this, it is evident that the classification of an asset as a CP24, CP34, or CP44 device is a structural prerequisite for participation in an I4.0 communication system. An asset of the CP-class 33 would be, for example, a field device that only has to record and provide data or fulfill a control task.

Taking this into account, every machine and every autonomous module can work as an I4.0-compliant object in the digital environment. If the perfect binder is used in the production of individual photo books, for example, it must comply with CP-class 44 and communicate via OPC UA with the production units involved and an ERP system[13] via its integrated

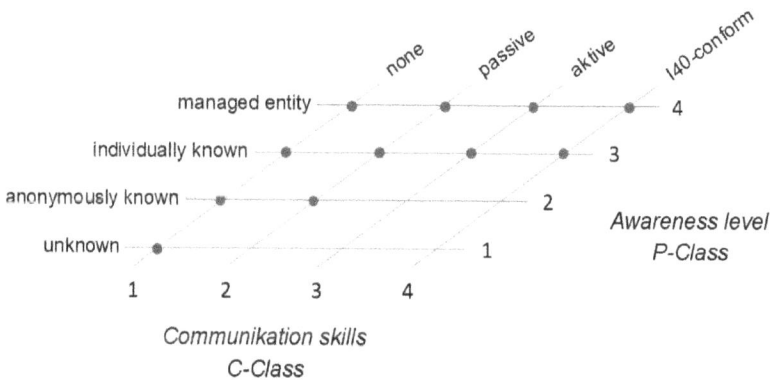

**Figure 2.12.** Illustration of the CP classification.

*Source:* According to [10].

---

[13] An ERP system (Enterprise Resource Planning) is used to manage business processes. This includes operational resources, such as capital, personnel, or means of production.

administration shell. The photo book itself would have CP23 if it has a QR code or RFID chip from which the production units can read the required technological parameters. A sensor for recording data for condition monitoring would have CP33 if it sends the data to a cloud via OPC UA.

### 2.2.4.4 *The administration shell*

The previous definitions clarify that the Asset Administration Shell (AAS) is a central component for the virtual and active representation of an I4.0 component [18]. The status report *structure of the administration shell — further development of the reference model for the Industry 4.0 component* [10] comprehensively describes the general requirements for an asset administration shell from the specifications of the *digital factory* and, in particular, its structure and design. In the following explanations, we only address the features of the asset administration shell that are relevant to machine and plant automation.

**Definition of the administration shell**
In [38], the asset administration shell is defined as follows:

> The asset administration shell (AAS) is an interoperable implementation of the digital twin and is standardized in IEC 63278. As a digital representation of an asset (e.g. device, machine, plant) over its entire life cycle, it acts as the *life cycle record* of an asset and is therefore the linchpin in the value creation system. An AAS can already exist before the actual asset has been developed or produced (e.g. as early as the quotation preparation stage). The AAS can also be used for assets that do not have a communication interface (e.g. screws).
>
> The most important features and roles defined for the asset administration shell:
>
> 1. An AAS has a reference to an asset and represents it digitally,
> 2. An AAS offers one or more interfaces,
> 3. An AAS references one or more submodels,
> 4. An AAS user application accesses the AAS information via data interface(s).[14]

---

[14]With regard to communication, it goes on to say: *(...) In the future, the AAS could be a direct have the ability to communicate. This is not yet fully specified, but is state of the art. Research and part of funded research projects.*

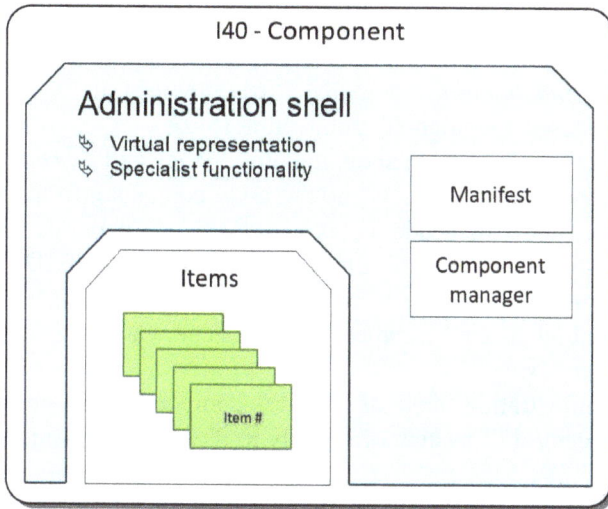

**Figure 2.13.** I4.0 component with an asset administration shell.

*Source*: According to [10].

The central task of an asset administration shell is therefore to create standardized digital access to information about the associated asset instance or asset type. In the process, associated asset information is electronically modeled via so-called submodels and can be found, explored, and used in a standardized way. Examples include the digital type of plate, geometry descriptions, device parameters, interfaces, documentation, etc.

In its function as the *central element of an I4.0 component,* the Administration Shell thus contains all the information for its *virtual representation* and the *technical functionality.*

Figure 2.13 symbolically shows the basic properties and components of the administration shell.

The *manifest* acts as a directory of the individual data content for *virtual representation.*[15] It also contains mandatory information on the I4.0 component, including the connection with the objects by means of a corresponding identification option and with the mechanisms for data security.

---

[15] See the comments on the significance of the *manifesto* in Section 2.2.3 independently of this, the content of the current status report "Structure of the Administration Shell" [10] without losing sight of future developments.

The *component manager* provides the connection to the IT services of the I4.0 component, which can be used to access the *virtual representation* and *technical functionality* externally. The component manager can therefore connect a service-oriented architecture (SOA).

The asset administration shell contains a *body* and a *header* and can refer to one or more objects. While the body contains information about the respective item, the header contains information about its use.

Each object, like the administration shell, has a unique ID.[16]

The identification of the administration shell in and its communication with the higher-level technical system occurs via a unique and standardized *semantics*.[17]

The administration shell of the I4.0 component can be physically located in a central IT system or directly in the technical device.[18]

**Structure of the administration shell**

The structure of the administration shell is based on established automation and information technology concepts that already exist and are used as standards in various industries. To fulfill its role, the asset administration shell records data and functions for a wide variety of application scenarios and makes them available to the surrounding technical system.

To the *characteristics* of the managed *device*[19] are assigned in *views*[20] to meet the different requirements of the environment and life cycles.[21] It is subject to the design decision for the specific component which feature exactly is assigned to which view. A feature can occur in several views. On the other hand, every characteristic that is relevant to the I4.0 environment must be assigned to at least one view. Table 2.2 lists defined *basic views* that were specified in [10] but which can be supplemented by additional views if required.

These basic features are supplemented by *mandatory features* that are mandatory and standardized for the submodels of the asset administration

---

[16] Cf. *primary key*, Section 2.2.1.

[17] The topic *Semantics* is discussed in detail in [10, 15] and is classified as a need for further research in [17]. Existing semantic descriptions can be found in [35].

[18] There are plans to integrate a management shell at chip level. This means that field devices can also interact directly as I4.0 components and simple integration into established control systems is not an insurmountable hurdle.

[19] Cf. entities and their properties, Section 2.2.1.

[20] Cf. balance sheet areas, Section 2.1.1.

[21] Cf. module property Encapsulation, Section 2.1.1.

**Table 2.2.** Overview of the defined basic views for the administration shell.

| Basic view | Best practice/examples |
|---|---|
| Business | Data and functions are deposited which allow judging on the business suitability and performance of a component in the life cycle phases Procurement, Design, Operation, and Realization. Examples: prices, terms of delivery, and order codes. |
| Constructive | Contains properties relevant for the constructive deployment of the component, thus for selection and building structure. |
| | Contains a structure classification system pursuant to EN 81346. |
| | Contains numerous properties with respect to physical dimensions and regarding start, processing, and output values of the component. |
| | Contains a modular view of subcomponents or a device structure. Allows an automation view with inputs and outputs of different signal types. |
| Performance | Describes performance and behavioral characteristics to allow a summary assessment and Virtual Commissioning (V-IBN) of an overall system. |
| Functional | Makes statements on the function pursuant to EN 81346 and on the function of the subcomponents. Here, location of the individual functions of the Technical Functionality also occurs, thus, for example, the so-called "skills", interpretation, commissioning, calculation, or diagnosis functions of the component. |
| Local | Makes statements on positions and local relationships between the component or its parts or inputs and outputs. |
| Security | Can identify a property as security relevant. This property should be considered for an assessment of security. |
| Network view | Makes statements in respect of electrical, fluidic, materials flow-related and logical cross-linking of the component. |
| Life cycle | Contains data on the current situation and historical utilization in the life cycle of the component. Examples: allocation to production, maintenance protocols, and past applications. |
| People | Properties, data, and functions should appear in all views so that humans can understand individual elements, inter-relationships. and causal chains. |

*Source:* According to [10].

shell [10]. There may also be *optional features* that are standardized but not mandatory. Finally, *free characteristics* are possible, e.g. containing manufacturer-specific information.[22]

In summary, the rough structure of the administration shell can be described as follows[23]:

- *The body* contains the *component manager*, which manages individual *submodels* within the administration shell. Each submodel has hierarchically organized features that reference *individual data and functions.*
- The entirety of all submodels forms the *manifest* as a clearly identifiable table of contents of all data and functions. The respective feature structures are available in a strict, standardized format.[24]
- The asset administration shell can record and display the object's *runtime data.*
- The information volumes are displayed externally by means of *views.*
- An *I4.0-compliant, service-oriented API* (Application Programmers Interface) makes the services of the component manager accessible to the outside world.
- The administration shell has exactly one *component manager.*[25] It can be implemented on any node in the network or, if the information storage and processing capacity of a physical asset allows it, on this asset itself. Integrated component managers enable self-administration and form the basis for CPS and IoT concepts.[26]

### 2.2.4.5  *Interaction of I4.0 components*

In the previous sections, we referenced the *service orientation* as the basic principle of the interaction of I4.0 components several times.[27] Figure 2.14 shows how I4.0-compliant technical devices interact exclusively via their administration shell. It must be ensured that the information exchanged

---

[22] See also further detailed descriptions of the features in [10].

[23] Cf. image and text on "Rough structure of the Administration shell" in [10].

[24] A structure in accordance with IEC 61360 [10] is recommended.

[25] The *component manager* is also referred to as the *resource manager* or *administrator* in various sources, although the same functionality is always assumed. functionality is always assumed.

[26] These implementation details are not specified by I4.0 and may vary from case to case.

[27] Cf. embassies and Services, Section 2.2.3.

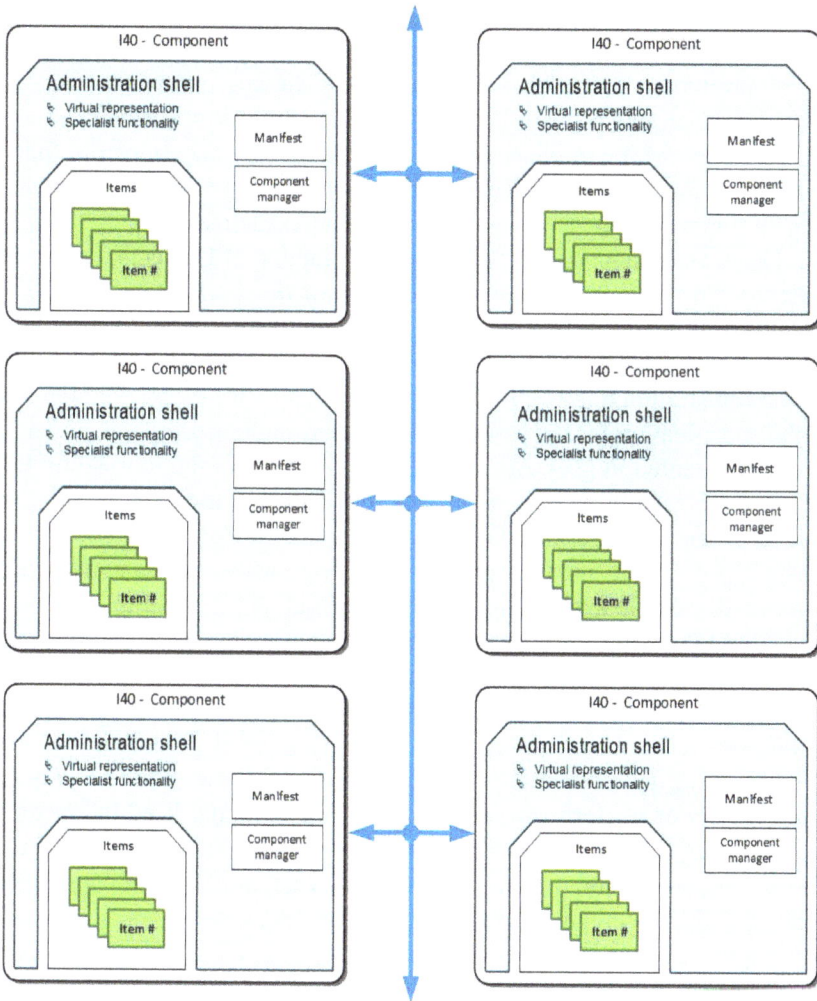

**Figure 2.14.** I4.0-compliant communication via the asset administration shell.

*Source*: According to [10].

meets the relevant IT security requirements. In addition, the exchange of operational data, such as the position of a robot's tool or a position-synchronized signal for image recognition, may be required in real time. It may also include information on machine safety, the service-oriented exchange of which generates additional requirements for engineering and safety-related approvals.

It is therefore part of the design decision of an asset or a production unit consisting of several assets, which components are integrated into digital production as an I4.0-compliant entity directly and which components are integrated as a sub-model within an administration shell.

However, I4.0-compliant interaction enables the access of the functionalities of technical assets by means of service access. Thus, it contributes an important part to industry-neutral data exchange.[28]

The development of a model for the interaction of I4.0 components is discussed in detail in *Further development of the interaction model for industry 4.0 components* [11] and is also included in *The Industry 4.0 research agenda* [17] and in the discussion paper — Interoperability with the Administration Shell ... [38]. Work on this is in full swing. Version 3.0 of the specification for the asset administration shell information model is already presented in [39]. According to IDTA,[29] it lays the foundation for the standardized digital twin in the industry. This specification defines the software structure, the interface, and the semantics of the AAS, which make it possible to provide data quickly and easily from an industrial plant to all participants along the value chain in an interoperable manner throughout the entire life cycle. The integration of an official interface to the AAS — the so-called API — is also important. This enables partners in the value chain to exchange their data across company boundaries via AAS [39]. It becomes apparent that service-oriented protocols such as OPC UA [20], the communication standard *PackML* [21], and *OOP* data types (JSON or rdf formats) form the basis for a standardized interaction model. Since the availability and distribution of OPC UA TSN, real-time interaction capability, including safety-relevant applications, has also been available for many production processes [22, 23, 37].

In [11], a further task of the administration shell is defined. It concerns the procedure for managing services based on *supply and demand*.[30] This means that an asset first asks a service provider whether the use of a required service offered by it is technically and organizationally possible. The use of a service is therefore negotiated beforehand. Figure 2.15 schematically shows this process based on the production level. In the scenario

---

[28] Cf. Compatibility, Section 2.1.4.

[29] Industrial Digital Twin Association (IDTA), www.industrialdigitaltwin.org.

[30] The issue of self-organization of production systems based on supply and demand is currently still at the research stage. Here too, opinions within the I4.0 committees differ somewhat.

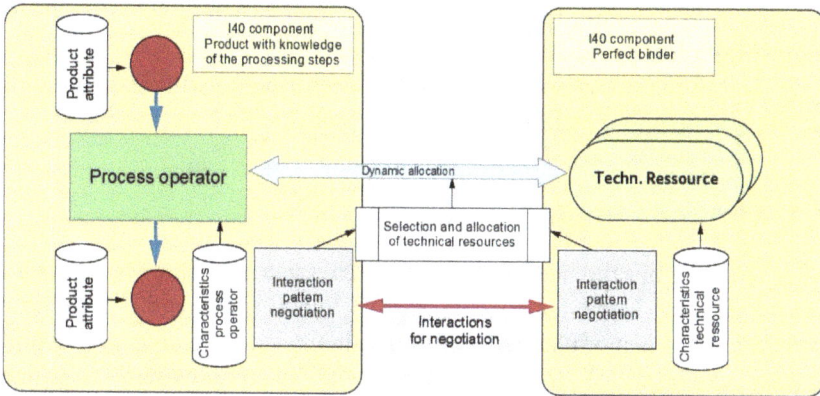

**Figure 2.15.** I4.0 components determine the assignment of process operator and technical resource by negotiation.

*Source*: According to [11].

presented, the product uses a process operator to ask the I4.0 component drilling machine whether it can be processed with the corresponding characteristics. If this is the case, the operator will allocate the resource and, if necessary, inform other components about it or provide further services, e.g. a transport order.

If we project this mode of action onto the *perfect binder*, then the transport system could ask the process operator (e.g. an ERP system) to which station it should transport the book block next after it has been transferred from the collating machine and identified via a QR code. In this example, the ERP system, as the higher-level I4.0 component, is aware of the processing steps and will negotiate with one of the existing perfect binders. The perfect binder may respond with the information that the variant of the gluing station required for the technologically necessary processing step is in a docking position, and it informs the customer that it will not be ready for operation for another 10 minutes. If this were the case, the ERP system could send the book block in question to a parking position, prompt the operating personnel with a message to retool, and then wait until the perfect binder has been retooled with an operational gluing station. Additionally, the ERP system could determine from its order queue whether there are other products to be processed using the same process and later guide them all one after the other to this perfect binder to minimize the changeover time.

Before we explain the definitions and aspects above in more detail using examples, we demonstrate which methods for modularization of machines and systems appear to be suitable for practical work. Many of the aspects mentioned above must also be considered here.

## 2.3 Methods of modularization

After working out the need for a *modularized machine and system struc-ture* in Chapter 1 and focusing on this in Sections 2.1 and 2.2 we now examine  how a modularized and individually configurable production system[31] can be created from — or instead of — a complex series machine or system. The literature offers a wide variety of design methods, depend-ing on the industry being considered and the objective being pursued. Many publications and standards relate to modularization in automotive engineering, both for the products and for the entire production and organization in its environment. The same applies to software engineer-ing, as the development of object-oriented structures requires a similar approach. We can learn from these specialist areas how modules must be designed to generate the expected benefits as such, but we can also learn something about engineering for the development of modular structures.

The publications by *TCW Transfer-Centrum für Produktions-Logistik und Technologie-Management GmbH & Co. KG, Munich,* and *ITQ-GmbH Garching* are of particular interest as well as the following selected sources:

- *Modularization in organization, products, production and service* [1]
  Basic statements on modularization strategies with a focus on automo-tive engineering.[32]
- *Concept for function-oriented systematic reuse in the engineering of automated systems in the process industry* [24]
  Engineering of automated systems based on reusable units.

---

[31] Unless there are explicit reasons for a distinction, we use the term "production system" for "machine or plant" in the following.

[32] The term "Modularization 4.0" is interesting in this work. The authors assure us that they only discovered it after they had decided on the title of their own book during their research. In fact, its approach harmonizes perfectly with the aim of "Automation 4.0".

- *Namur Recommendation 148: Requirements for automation technology due to the modularization of process plants* [5]
  Recommendations for modular plant engineering with a focus on automation technology.
- *Lehr- und Übungsbuch der Informatik, Band 3* [14]
  The techniques described for developing high-quality software, such as object-oriented analysis (OOA), object-oriented design (OOE), and object-oriented design (OOD), are not only very suitable for software engineering but can also be seen as a model for modularization in general.

We want to apply this knowledge and limit our further considerations of modularization to the automation aspects of a modularized production system. Therefore, we do not consider the monetary evaluation of modular concepts at this point. Although — and this should be expressly emphasized — this is an essential part of a complete modularization process. The same applies to the accompanying organization of production and service. The following section can only represent a small part of the problem.

## 2.3.1 Established design methods

If a modular structure should be created, the main task is to analyze all the requirements for a system step by step. This is divided into two main methods, the *top-down* and the *bottom-up method.*

In the *top-down method*, the hierarchical division of an overall system into its individual components takes place in what is known as a step-by-step decomposition. For this purpose, smaller and smaller balance spaces are defined within the complex structure until so-called *atomic subsystems* are created that can no longer be meaningfully decomposed. In this way, we move from an overall structure to the details. The result is a model of the modular overall system containing modules with their functions, interfaces, and associations.

The *bottom-up method,* on the other hand, starts by first defining the sub-elements with their interactions and methods, and the overall system is created step by step. The danger with this method is that the overall structure only emerges gradually. Knowledge that is only gained at a late stage must be incorporated iteration by iteration into existing module definitions. For this reason, this so-called step-by-step aggregation is only sensible if partial elements already exist and are to be reused or existing overall concepts are to be expanded.

In practice, both methods can be justified. However, the *top-down method* certainly appears to be the most suitable for the initial analysis of existing systems, as it is based on the realization that humans cannot take on more than seven to nine problems at the same time. In addition, humans perceive a large coherent problem as larger than the sum of the complexities of its individual parts. However, this problem can be tackled with the principle of "divide and conquer". By successively breaking down the large problem into smaller and more manageable tasks, it is possible to solve the subtasks without difficulty [13]. As already mentioned in Section 2.1.1, the most important task is to *define the balance areas*. The criteria according to which this is done solely depends on the type of system to be converted into a modular structure and the objective. If, for example, the aim is to divide a rather voluminous machine into suitable transport units, *design* criteria will probably be of more importance. If, on the other hand, an individualized series machine is to be created, the *functional* arguments will be more important. In principle, however, it can be assumed that a module must be defined in terms of both design and process engineering functions. This is why every modularization begins with a comprehensive analysis of the actual state and the definition of the target state, the results of which are presented in the following models[33] and can thus be incorporated directly into a specification.[34]

This analysis results in the following approaches based on the *Unified Modelling-Language (UML)*[35]:

- **Function diagram**
  Determination of the necessary and expected functions and presentation of their sequence for processing the product or product type under consideration.
- **Class diagram**
  Overview of the object types recognized in the system, their external functionalities, and logical relationships to each other.

---

[33] Based on the method of Booch [30].
[34] See Chapter 3.
[35] UML is specified in the ISO/IEC 19505 standard and is the standard description tool for software modeling.

- **State diagram**
  Representation of the dynamic behavior of an object type.[36]
- **Interaction diagram**
  Representation of the interaction of the object types with each other and of the overall system with the environment.
- **Module diagram**
  Representation of the physical distribution of the individual object types.

In the following section, we take a closer look at the individual work steps.

## 2.3.2  Target analysis of the requirements

There is an old saying that "All beginnings are difficult". This certainly applies to the design of any technical device and even more so to the design of an individually configurable production system. But in this case, there is an extremely logical starting point — the view of the *product* to be processed or the *product* being created.[37] This is based on the realization that all conceivable production systems only fulfill a single purpose, namely, to produce something. Whether a machine, a plant, a power station, or an oven, they all manufacture products. The perfect binder produces a book, the refinery processes crude oil into fuel, the power plant generates electrical energy, and an oven bakes bread. So, what could be more obvious than to start by looking at the product in question?

In the following sections, we want to present some of the most important criteria with their relevance for the entire design and explicitly for automation, on the basis of which the initial and basic requirements for the production system to be developed can be determined.

In Figure 2.16, it becomes apparent that requirements analysis is a team effort across different disciplines, involving technologists and designers as well as employees from product marketing and sales. And because priorities and responsibilities can vary greatly across companies and from product to product, the following explanations deliberately

---

[36] Cf. Section 2.1.3.

[37] Since the production system is also a product, we use the term "end product" where necessary to make a clearer distinction, meaning the product to be manufactured in the production system.

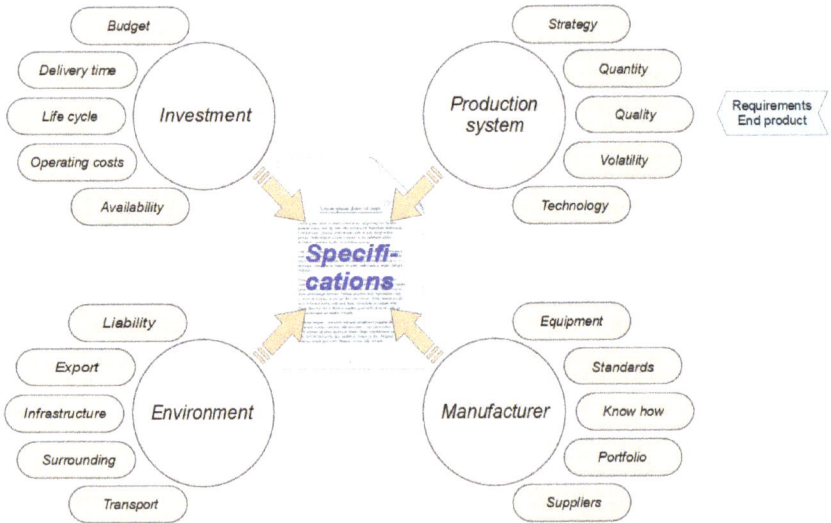

**Figure 2.16.** Overview of the requirements for the design specifications.

refrain from assigning responsibilities and presenting an order of priority, nor do they claim to be exhaustive. As a product, we want to start from a production system to be manufactured and directly point out the conse- quences for the design process and the design of the automation at each point (Figures 2.17–2.20).

### 2.3.2.1 *Product view*

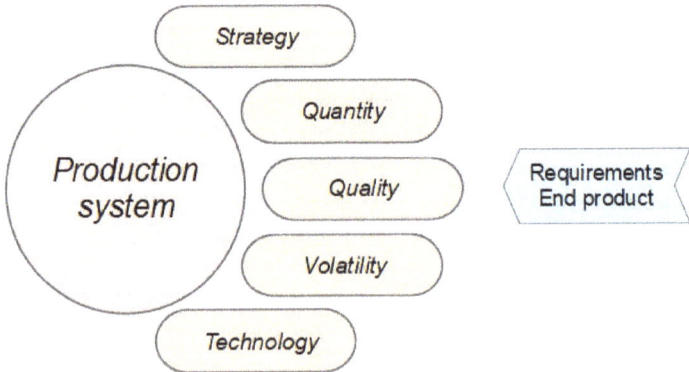

**Figure 2.17.** Criteria for the requirements analysis from the perspective of product.

**Target price of the product**
This point is certainly one of or even the most important when placing any product on the market, as it ultimately plays a key role above all other aspects. Regardless of whether the intended quality, the material to be used, the required technology, or the number of units to be produced per unit of time, all other decisions are derived from the market price that can or must be achieved with the product.

## Construction

- Price sensitivity throughout the entire development and manufacturing process of the production system
- Reducing investment costs through modularization and reuse strategies
- Ensuring maximum effectiveness using the production system

## Automation 4.0 Perspective

- Shifting mechanical expertise to software to reduce production costs
- Provide equipment for automated conversion
- Offering intelligent service

**Quantity (products per time unit)**
The quantity to be produced per unit of time is a decisive benchmark for the profitability of production systems. The expected product output is decisive for the selection of technological process steps, material supply and disposal, and many other aspects.

## Construction

- Observe targets and exceed them if possible
- High speeds also mean higher stress on the mechanics

**Automation 4.0 Perspective**

- Powerful control and drive technology
- Higher requirements for the design of safety technology
- Provide condition monitoring for highly loaded drives to avoid unforeseen downtimes

## Quality features

This is not just about the quality of the product itself but also about safeguarding it. In the case of production systems for the food or pharmaceutical industries, for example, appropriate quality assurance aspects must be incorporated into the design decisions from the outset.

**Construction**

- Enable the best possible reproduction rate through low tolerances

**Automation 4.0 Perspective**

- Provide equipment for integrated quality assurance
- Quality data achieved must be saved or passed on

## Volatility (variety of variants)

What is the life cycle of the product? Which product variety and types should be able to be manufactured on the production system? Is it always the same mass-produced product or an individual series product and what changes are to be expected? Will different materials be used?

Answering these questions is crucial for the functional and constructive design of the production system.

## ✓ Construction

- High volatility forces intelligent conversion
- Enable adaptation of the production system through a modular concept

## ⮕ Automation 4.0 Perspective

- Provide recipe system and equipment for automated changeover
- Mechanical module concepts require modular automation technology
- Integration into digital production if necessary

**Manufacturing technology**

This analysis considers all aspects of product manufacture. This also includes recording the required raw and auxiliary materials, the individual process steps, the knowledge and treatment of waste products, and the energy requirements, such as the electrical energy, compressed air, or special gases required for production. In the case of individual series products, several technologies may need to be implemented.

## ✓ Construction

- Use of new technologies means additional effort throughout the entire life cycle of the production system (engineering, production, training, service, etc.)
- Using proven technologies and processes where possible

## ⮕ Automation 4.0 Perspective

- If possible, use technology packages from the manufacturer
- Integration of external devices into the automation system (operational data, operation, configuration, diagnostics, etc.)

## 2.3.2.2 *Investment view*

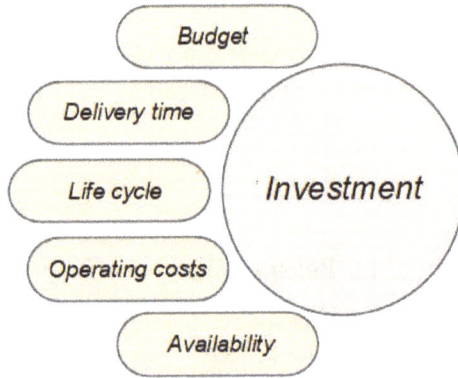

**Figure 2.18.**   Criteria for the requirements analysis from the perspective of investment.

**Investment budget**

Sometimes the customer or the market determines the costs for investing in a production system t at a very early stage. Regardless, most decisions will be based on the investment budget. After all, a production system itself is a product and is in principle subject to the same requirements as the product to be manufactured.

## ✓ Construction

- Price sensitivity throughout the entire development and manufacturing process of the production system
- Reducing investment costs through modularization and reuse strategies

## ⇨ Automation 4.0 Perspective

- Shifting mechanical expertise to software to reduce production costs
- Generate added value for the production system through additional functions and appealing operating concepts

**Delivery time and preparation for use**

The time from placing the order to the ready-to-use handover of the production system to the customer is a binding part of every supply contract and must be adhered to. Delays often result in contractual penalties and not only does the supplier's image suffer but so does its margin. The shortest possible delivery times, on the other hand, enhance the image and ensure the effectiveness of the manufacturer. The instruction and training of the operating and service personnel is crucial, ensuring that the equipment is ready for use.

✓ **Construction**

- Reducing delivery times and making them predictable through modularization and reuse strategies
- Reducing dependence on suppliers through in-house expertise

⇨ **Automation 4.0 Perspective**

- Early simulation and high software quality avoid surprises in production and commissioning
- Avoiding new developments through modularization and reuse of hardware and software

**Life cycle**

A production system is subject to aging processes and changes. Maintenance and changeover costs for changing products must therefore be considered from the outset. After all, it also has a significant impact on the cumulative cash flow (CCF).

✓ **Construction**

- Ensuring the sustainability of the investment through adaptable design
- Analyze future developments of the end product at an early stage and enable retrofitting through a modular concept
- Modularity ensures that system components can be replaced in the event of wear or new technology

**Automation 4.0 Perspective**

- Provide condition monitoring for critical components
- Limit the effort required for changes through modular hardware and software design

**Operating costs**

This aspect is an important decision criterion for the overall *equipment effectiveness (OEE)* of the operator. Not only do the costs for the required materials, energy, or personnel play a role here. Downtimes due to retooling or breakdowns also cause additional costs due to loss of production.

**Construction**

- Using new technological possibilities to reduce material, energy, and time consumption
- Enable resource-saving operating modes, e.g. standby mode
- Minimize maintenance costs with durable and maintenance-free components

**Automation 4.0 Perspective**

- New technological possibilities require innovative control and regulation processes
- Use energy-saving components

**Availability**

One should not underestimate the operating mode of a production system. Should production take place 365 days a year and around the clock or only in a single-shift system? Does a shutdown of just this production system also initiate a shutdown of the plant's entire production? These criteria determine which equipment and spare parts need to be kept in stock either by the supplier or the operator, whether and which resources need to be used for condition monitoring, or whether automated retooling is required for highly volatile products.

## ✓ Construction

- Provide equipment for automated conversion
- Increase reliability with durable and maintenance-free components

## ⇥ Automation 4.0 Perspective

- Enable effective operation and avoid operating errors through intelligent concepts
- Provide condition monitoring for critical components
- Offer intelligent service

### 2.3.2.3 *Production environment*

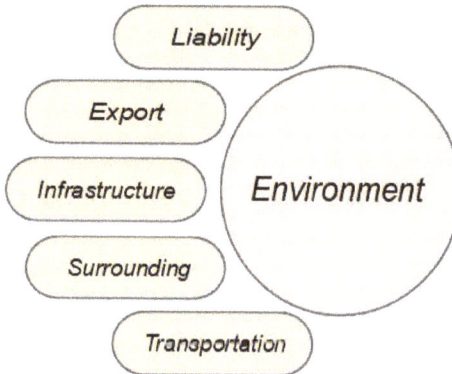

**Figure 2.19.** Criteria for the requirements analysis from the perspective of production environment.

## Product liability

This aspect affects both the end product and the production system. Both must be offered and operated in compliance with standards and legislation. Export-oriented manufacturers must comply with many international and national regulations. In addition, there are different opinions on the legal assessment of a loss of production, an accident at work, or a single or systematic deviation in quality. This also applies to the protection of your own know-how.

### ✓ Construction

- Compliance with all standards, regulations, and laws
- Careful and legally compliant development, production, and documentation

### Automation 4.0 Perspective

- Possibly increased effort for the collection, storage, and forwarding of quality data
- Certain programming languages and forms of documentation may be required
- Measures to protect know-how

**Export regulations**

Unfortunately, there are numerous customs barriers and embargo regulations for capital goods worldwide, which must be considered during the engineering process.

### ✓ Construction

- Legally compliant selection of individual components and technologies

### Automation 4.0 Perspective

- Often, only automation and information technology products are affected by embargoes, meaning that certain technological requirements cannot be met in some cases

**Infrastructure**

The supply of materials and media, such as gases, fluids, or compressed air, is generally a technological requirement and does not pose a particular problem if the supply contract is structured accordingly. After all, the system operator is interested in and responsible for their uninterrupted supply. If production systems are also to be integrated into digital production or remote maintenance is to be used, the respective system must be given access to the IT infrastructure. This is also at the discretion and responsibility of the operator, provided that the manufacturer provides the corresponding interfaces on the production system and the interaction has been carried out as agreed. Usually, the operator is also responsible for IT security, although the boundaries here are somewhat more blurry. Manufacturers of production systems are therefore well advised to provide their own measures, such as a firewall in hardware and/or software.[38]

The situation is different when it comes to the supply of electrical energy. It is not uncommon in many regions for sudden power failures to appear leading to serious damage in production facilities and almost finished products if the appropriate precautions are not taken.

✓ **Construction**

- Clarification of the media supply
- Take constructive precautions to prevent damage to the production system and end product in case of a sudden power failure

**Automation 4.0 Perspective**

- Appropriate monitoring of the technological media supply using suitable sensors/actuators
- Provide suitable interfaces, communication protocols, and data security measures for integration into the IT infrastructure
- In case of a sudden power failure, machines with moving axes can be brought to a standstill in a targeted and technology-compliant manner by recovering mechanical energy

---

[38] See Section 4.4.5.

**Surroundings**

This point includes all boundary conditions relating to the immediate production environment. First, this includes all information about the installation site. It must be clarified whether the production system will be in an air-conditioned factory hall, for example, or whether it will be operated in an open-air facility. This is followed by information on ambient temperatures, humidity, floor load-bearing capacity, and vibrations and, especially in chemical and process engineering systems, consideration of hazards from escaping substances and explosion protection requirements. The availability and qualification of the intended operating and service personnel must also be considered at this point.

> ✓ **Construction**
>
> - Precise recording of the ambient conditions and correspondingly adapted design

> ⮕ **Automation 4.0 Perspective**
>
> - Careful device selection and special measures to protect against high temperatures, moisture, dust, corrosion, and explosion
> - Development of suitable operating concepts

**Transportation**

It may come as a surprise, but the transportation of a production system from the manufacturer to the operator must also be considered in the design. It makes a difference whether the rather small machine fits into a box and can be transported anywhere in the world by any carrier or whether a bulky production system must first be assembled by the manufacturer, verified by the customer, and then sent on its way partially disassembled. Suitable designs and scenarios must be developed to avoid straining the assembly and commissioning effort at the destination. The shipping routes must also be taken into account (truck, ship, flight, etc.), and suitable packaging, transport safety devices, and suitable lifting equipment must be provided.

☑ **Construction**

- Incorporate transportation and shipping options into the design
- This is another reason for creating modular structures

↪ **Automation 4.0 Perspective**

- Transportation and shipping options can also influence device selection
- Observe the conditions for assembly and transport-related disassembly
- Enable commissioning in general and especially on site without development personnel

### 2.3.2.4 *Manufacturer's view*

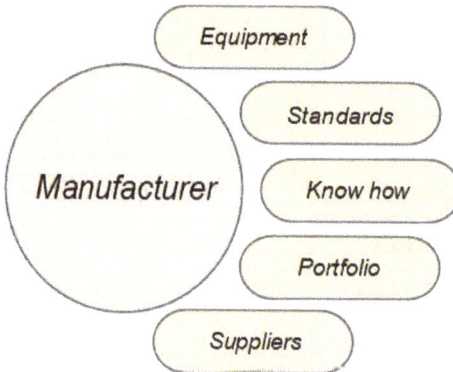

**Figure 2.20.** Criteria for the requirements analysis from the perspective of manufacturer.

**Available production equipment**

It goes without saying that the manufacturer's production capacity and equipment, including machinery and the corresponding specialist personnel, must be incorporated into every system concept. However, with increasing technological requirements, a company's own added value sometimes reaches its limits more quickly than expected. External resources must be used to integrate certain processes into the company's own system. It is easily underestimated that the production of a module or a more complex component must be coordinated in a way that these subsystems can then be integrated into the overall system without any repercussions and without additional effort.

### Construction

- Precise design and functional specifications are required for delivery

### Automation 4.0 Perspective

- Develop precise technical and functional specifications for the delivery

**Existing company standards**

In addition to external regulations and standards, there are usually also internal company standards that must also be taken into account.

### Construction

- Development according to internal company specifications

### Automation 4.0 Perspective

- Development according to internal company specifications

**Company know-how**

This is similar to the production equipment. Clear structuring and modularization of development and manufacturing processes can make the manufacture of a production system less susceptible to unexpected failure or partial lack of know-how. Development and production processes can also be structured in a modular way if work is carried out in specialist departments and project groups with the appropriate specialist personnel.

### ✓ Construction

- Modularizing the development and production process
- Define clear boundaries and responsibilities

### ⮐ Automation 4.0 Perspective

- Ongoing training of development staff is required due to their importance
- Draw on manufacturers' know-how and technology packages without generating unnecessary dependencies

**Product portfolio**

When looking at a manufacturer's product portfolio, synergies can be identified that can be incorporated into development. It may also be possible to use components from other products (platform concept).

### ✓ Construction

- Use reusable units and solutions where possible
- Company standards must be strictly adhered to in every direction

### ⮐ Automation 4.0 Perspective

- Identical to the construction

**Supplier relationships**

We have already addressed the issue of selecting and involving suppliers at the beginning. At this point, it must be evaluated to which extent the material and immaterial supplier relationships match the requirements for the investment and the future location of the production system. If, for example, products are exported worldwide, the supplier must also be able to provide worldwide service under the same conditions. The same applies to the supplier's innovation behavior. The supplier must guarantee that replacement devices can be provided throughout the entire life cycle with the same external behavior, i.e. that they are structurally and functionally compatible. Innovations, e.g. by discontinuing components or eliminating faults, must be possible.

## Construction

- Anchoring the constructive and functional properties in the contract design

## Automation 4.0 Perspective

- Identical to the construction

### 2.3.3  Constructive detailed analysis

The results from the analysis of all requirements can now be presented as specifications for the design in a specification sheet. Figure 2.21 shows a selection of the criteria relevant to automation that need to be analyzed in detail in the further development process. In the following, we explain these in detail from the point of view of automation.

**Technology**

The first step is to decide on the definition and selection of the required production technologies. Even if these have already been largely fixed with the product view, detailed planning must take place at this point. This

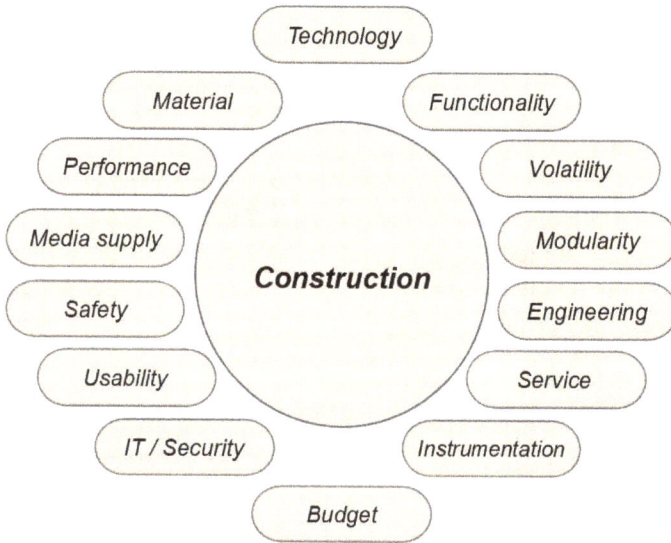

**Figure 2.21.** Criteria for the detailed design analysis of a production system.

results in the basic design of the production system and specifications for further mechanical and automation development.

## Automation 4.0 Perspective

- Introduction of automation technology packages to achieve a sensible reduction in mechanical components

### Functionality

Defining the technological functions of the production system is crucial for the entire subsequent development and design process. According to the method of gradual refinement, the internal functions are successively separated and specified based on the external functions in relation to the product.

## Automation 4.0 Perspective

- The resulting functional model is the basis for hardware and software development

**Volatility**

The changeability of the product is also determined by the required volatility of the production system. The necessary changeover scenarios must be defined in this design step, especially with regard to the aspects of product quantity and system availability. One must distinguish between automated changeovers (e.g. various format settings) and those that must be carried out by operating personnel. This is necessary if further technological changes are required for certain production steps (e.g. the replacement of production units). The results are incorporated into the rest of the development and design process.

## Automation 4.0 Perspective

- Price-sensitive selection of the hardware required for automated retrofitting
- Definition of interaction and data handling requirements
- Define I4.0 conformity when operating in a digital production environment

**Modularity**

The design for a modular structure of the production system requires corresponding results and decisions from the previous analysis steps, particularly the volatility. In this step, all technological and constructive units are defined with the corresponding properties and interactions. Existing technology modules are checked for their suitability and integrated into a modular concept. Only then is the design for the necessary new mechatronic units carried out.

## Automation 4.0 Perspective

- Design and further development of a modular hardware and software concept
- Decentralized instrumentation of drives, safety components, and operating units
- Mechatronic units from suppliers require special consideration of the necessary real-time, compatibility, and interaction requirements

## Engineering

Selection and definition of the project teams and the necessary tools, considering the existing and externally available know-how.[39]

### Automation 4.0 Perspective

- Selection of suitable tools for electrical design and software development
- Shortening development times through early simulation

## Material

It is important to know which materials and auxiliary materials are required for the selected technologies in order to plan the corresponding supply and disposal systems. This includes, for example, handling systems and conveying equipment, but also pipelines with valves, valves, pumps, and much more. The appropriate measures for quality control of the raw materials must also be defined.

### Automation 4.0 Perspective

- Upstream and downstream systems may be suppliers and must be integrated into the control of the production system via appropriate interfaces

## Performance

As a consequence of the required quantity, the dynamic requirements for all components of the production system are defined.

### Automation 4.0 Perspective

- Determining the requirements for the real-time capability of the automation system

---

[39] See Chapter 3 "Project planning for modular machines".

**Media**

At this point, the supply of the production system with electrical energy and other energy sources such as compressed air or hydraulics is planned.

## Automation 4.0 Perspective

- Determination of the requirements for the necessary measurement and control technology for quantity recording, dosing, and monitoring

**Safety**

It is crucial to carry out a risk analysis for the entire production system to rule out hazards for people, production facilities, products, and the environment. Both mechanical and automation measures to minimize risks must then be defined in a standardized process.[40]

## Automation 4.0 Perspective

- Eliminate the motivation to manipulate safety devices with intelligent safety technology while also avoiding costly design measures at the same time
- Intelligent security technology enables new technological processes and functions

**Usability**

This process step includes the definition of the operating concept for the entire production system and, if necessary, other decentralized operating stations and elements. This also includes defining the user roles and rights required for operation as well as the options for remote access.

## Automation 4.0 Perspective

- Integrate the possibilities of modern, effective, and error-proof human–machine communication into the concept
- Define interactions with integrated and higher-level systems required for operation

---

[40] A detailed description of this process can be found in [29].

## IT/Security

Development of a concept to protect the production system against deliberate manipulation by humans. This includes not only data security when integrated into an IT system but also the handling of various accessible interfaces such as USB or LAN and other interfaces.

### Automation 4.0 Perspective

- Introduce possibilities for effective security measures and mechanisms

## Service

Appropriate specifications must be made at an early stage of development to ensure effective servicing. This includes suitable design measures, such as easily detachable and easily accessible connection technology without the need for special tools, as well as preventive maintenance measures and easy-to-understand operating concepts.

### Automation 4.0 Perspective

- Preventive maintenance with condition monitoring
- Support concept for critical components
- Enable clear and intuitive operation of the entire production system
- Create meaningful documentation and make it available in the operating system
- Guarantee of remote diagnosis by the manufacturer's and operator's service personnel

## Instrumentation

The subcomponents can be designed and the required equipment technology selected in accordance with the detailed analysis to date.

### Automation 4.0 Perspective

- Selection of hardware and software components in line with requirements

**Budget**
Ongoing monitoring of budget targets during the entire development process.

> ### Automation 4.0 Perspective
>
> - Ensure range of functionality through high software content
> - Avoid unnecessary assemblies such as adapters, bridges, or measurement and control technology by selecting suitable technological components and software-based technology packages
> - Early and consistent use of simulation methods and innovative design tools

## 2.4 Modelling

The previous sections presented the most important methods, fundamentals, and means of description for the design process of modular production systems. We explained the basic properties of modules in the form of mechatronic units and introduced an object-oriented representation and description method. The most important requirements were analyzed to be able to meet the requirements of the necessary I4.0 conformity soon. Moreover, solutions that are already standardized or under development were presented. Finally, a methodical approach to the development of modular production systems was presented focusing on the requirements of future-proof automation systems. But how does this lead to the development of a modular model for a production system?

Basically, the modeling process begins again with the question of the causal goal for the modular design. If the goal of modularization is to break down a large production system into manageable parts solely for the purpose of easier and more efficient transport, mechanical aspects are more likely to play a significant role. However, when it comes to designing a flexible and customizable production system in the context of digital production, functional arguments are more important. To address the purpose of this book, we will examine this aspect more closely. To do this, we will again use the example of a *perfect binder*. But it should be considered that, in the interests of general understanding, this can only be a highly simplified excerpt of the design process. A complete description would go beyond the scope of this book.

Perfect binder

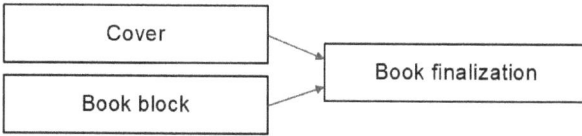

**Figure 2.22.** Illustration of the basic functions for a perfect binder.

## 2.4.1 Design of a functional structure

As already mentioned in Section 2.3.3 under "Functionality", the *functional analysis fundamentally* determines the entire development and design process. The aim of this analysis is to model a functional structure, whereby the individual functions are represented as abstractly as possible. The design process therefore starts with the analysis of the technological process and the definition of the required functions.

### 2.4.1.1 *The function and class diagram*

If the technological functions are shown in their logical sequence, an initial *function diagram*[41] of the production system is created. Figure 2.22 shows the technological process for the example of the perfect binder. This consists of three sub-processes: *book block processing, cover feeding,* and *finalization*. This diagram already shows that the book block *processing* and *binding* processes can run in parallel.

The process steps are then broken down further into their components, resulting in a diagram with the necessary basic functions and all optional functions. Figure 2.23 shows the result for the perfect binder.

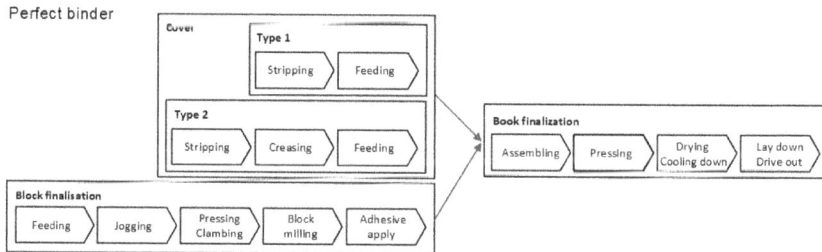

**Figure 2.23.** Functional diagram for a perfect binder.

---

[41] Cf. Section 2.3.1.

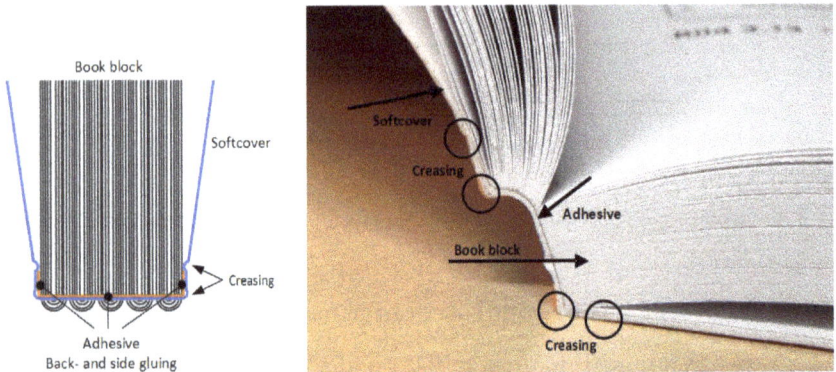

**Figure 2.24.**   Perfect binding of a paperback with 4-fold creased cover and block pages glued together.

During book block processing, the functions of *feeding* the book block, *jogging* (for a homogeneous block structure), *pressing* (so that the air escapes from the block), and *clambing* (to fix the aligned book block for the further processing step) are added. In the *spine processing* step, the gutters of the folded individual sheets are completely or partially cut open with a rotating *milling machine*, and *adhesive* is then applied to the spine and pages of the block in the gluing station. Depending on the required binding technology, further processing steps are required before the prepared book block reaches the finalization stage. At the same time, the *cover* is *removed* from a stack and prepared for the subsequent process if required. For example, a softcover binding still needs to be *creased* for precise folding and better opening behavior of the finished book. The cover is then sent to finalization, where it is first *placed* on the prepared book block and *aligned*.[42] The cover is then *pressed onto* the spine and pages, and the book must then remain in this state for a certain amount of time to allow the adhesive to *cool* and *dry* to achieve the minimum adhesion required for further transportation. The length of the dwell time depends on the gluing technology (type and quantity of adhesive) before it is *laid down* and *transported*.

The final step in the production of a brochure is three-side trimming, in which the finished products are given their final shape. However, the three-side trimmer required for this is a separate machine and is therefore not considered any further. Figure 2.24 shows the perfect binding for a

---

[42] We use the term *merging to* summarize these two process steps.

brochure with softcover as a 4-fold creased cover and block page gluing before and after trimming.

We already recognize a special case in the *binding* process. It would of course be possible to represent this as a single block, as from a functional perspective, what lies behind it must be irrelevant. However, two additional facts should be presented here. First, a book production always requires a cover. In our example, this should be different types such as a simple soft cover and a special cardboard. There are therefore two options for this sub-process. It is shown that the feeding of a soft cover with creasing contains an additional process step. For the rest of the illustration, we initially assume that these two variants are designed as derivatives and are option-ally inserted into the machine depending on the product requirements.

Furthermore, we assume that, depending on these two options, the spine processing, adhesive application, and joining processes must also have different characteristics and that additional work steps are also required for certain adhesive binding processes before joining. For exam-ple, for a flex-stable binding,[43] a paper strip must also be glued over the milled area in an intermediate step, followed by a further application of adhesive. This means that derivatives are required for some processes. Figure 2.25 shows these relationships (possible options are shaded).

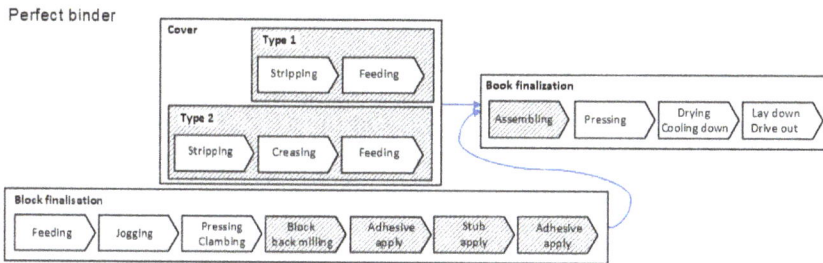

**Figure 2.25.** Function diagram for a perfect binder showing the basic functions and selected options.

Before we present these basic functions in further steps with the inter-nal functions, we must clarify further considerations regarding the sequence of the technology steps, as the previous flowcharts do not yet contain any functions for the auxiliary processes. In Figure 2.26 the sup-ply of electrical energy, suction and compressed air and material transport

---

[43] See the description of different industrial bookbinding processes in [34] or [33].

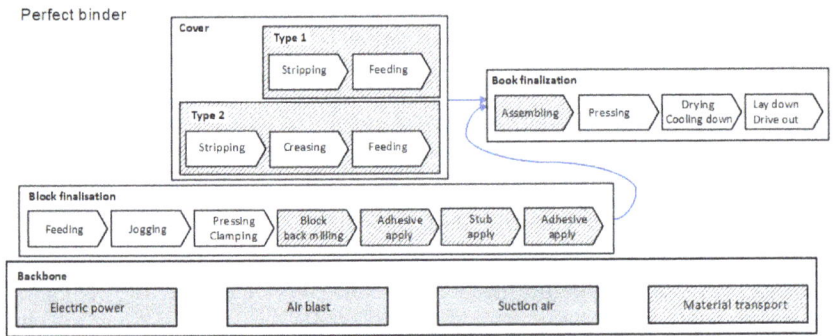

**Figure 2.26.**  Further developed function diagram for a perfect binder with additional representation of the auxiliary functions.

are shown as examples of secondary functions. The specific basic functions that must access these processes will be identified later in the corresponding ERD diagrams.

In the further *functional decomposition,* all basic functions are now broken-down step by step into their *atomic components* and the inner functions with their properties and relationships are thus successively defined. As we are developing a modular production system, the individual basic functions can already be regarded as entities for the representation in an ERD. In Figure 2.27, the inner functions are now entered under the outer function *Block back milling,* and Figure 2.28 shows the associated ERD as part of the *class diagram.*

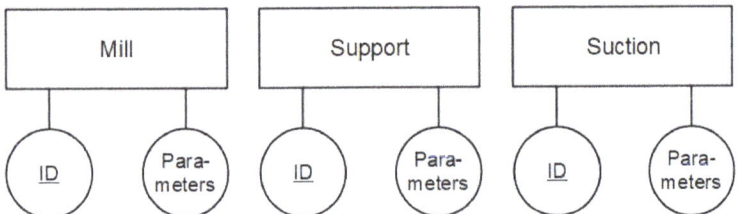

**Figure 2.27.**  Illustration of the internal functionalities of the mechatronic back processing.

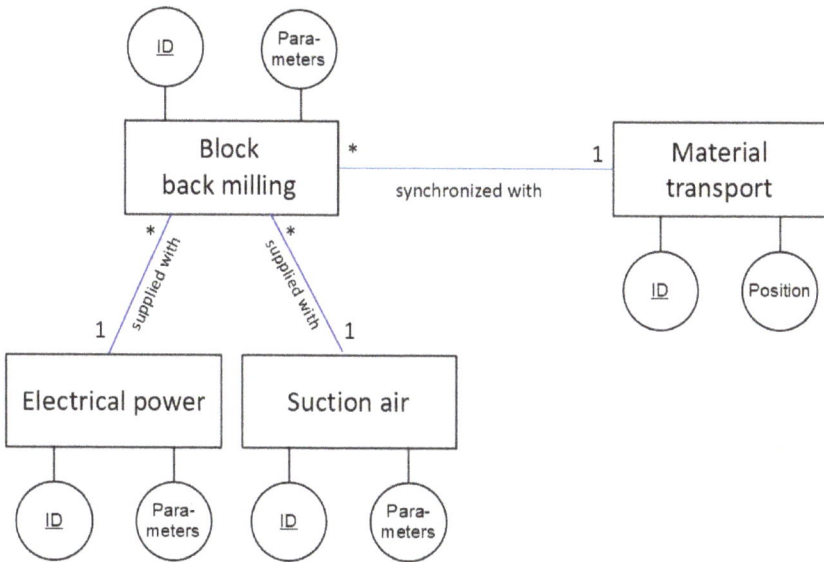

**Figure 2.28.** Simplified ERD of the mechatronic back machining component.

### 2.4.1.2 *The state diagram*

Once all functional components have been determined in our example, the methods can now be defined and displayed in a *state diagram*. This starts again with the external functions using the *top-down method*.

The milling process will be explained in more detail at this point for a better understanding (Figure 2.29). The milling head is in the *OFF* and *energy-saving* states in the *rest* position. To enable optimized cycle times in the normal production process, the milling head should be in the *ready* state, positioned to reach the milling position with minimal time and effort. This is directly below the reference position *h*. When the front edge of a book block reaches the processing station, a trigger signal is generated by a sensor and the milling process starts synchronously with the speed of the transport system. Depending on the type of binding process, the milling head is moved to the *working* position in accordance with the technological parameters $a_1$ and Z. With the movement of the transport system, the book spine is now milled at a set speed $\omega$ until the milling head swivels back into the *ready* position depending on parameter $a_2$. Alternatively, the milling head can also remain constantly in the *working* position. In this case, the entire spine of the book is always milled, as is typical for the production of paperbacks.

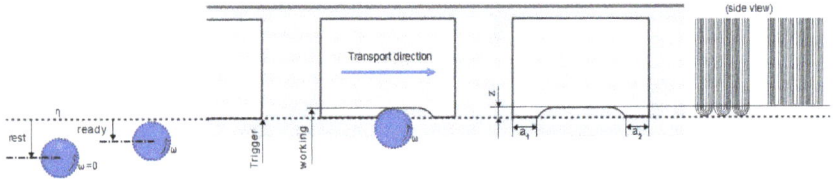

**Figure 2.29.** Technological sequence of the milling process using the example of the flex-stable bond.

The exact values for the swiveling in and out of the milling head result from the parameters $a_1$ and $a_2$, the possible infeed speed of the milling head, and the transport speed of the book block.[44] Table 2.3 and Figure 2.30 show the individual states with their properties for controlling the *back machining* from the perspective of the *perfect binder* in a simplified form.

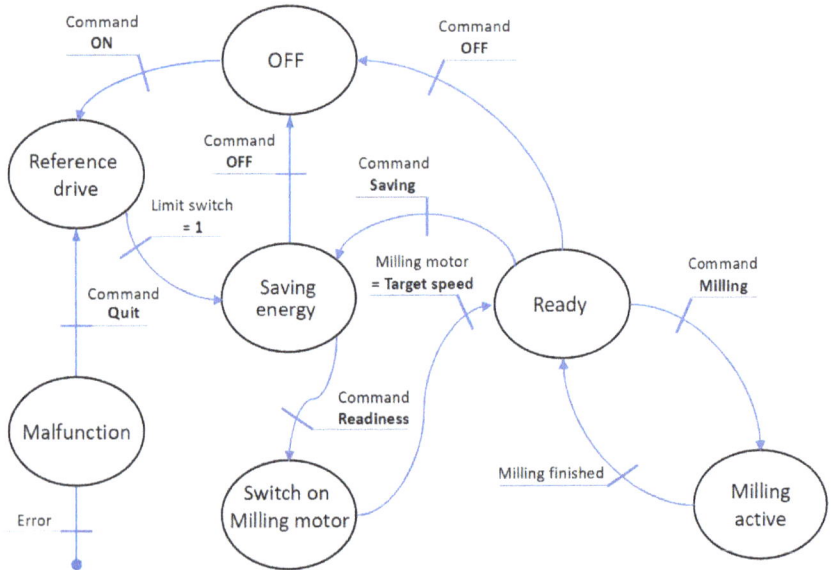

**Figure 2.30.** Simplified Moore automat for the back processing module.

[44] In practice, either the technology of the *dynamic dead time compensated cam controller* or, if the milling head infeed is a servo drive, an *electronic cam disk is* used to calculate and control this process.

**Table 2.3.** Overview of the states and actions of the back processing module.

| Condition | Inputs | | | | Outputs | | | Actions | | Transition condition | Subsequent state |
|---|---|---|---|---|---|---|---|---|---|---|---|
| | Position milling machine | Position master | Trigger signal | Limit switch | Speed milling motor | Feeding motor | Milling motor | Entrance | Output | | |
| OFF | x | x | x | x | x | 0 | 0 | Milling motor = off Infeed motor = off Suction = off | x | Command **ON** | Reference-ride |
| Reference-ride | x | x | x | = 1? | = 0 | Drive negative | 0 | Feeding motor backwards | Feed motor off | Limit switch = 1 | Energy save |
| Energy save | = *rest* | x | x | 1 | x | x | 0 | Milling motor = off Move milling machine to position *rest* | x | Command **OFF** — Command **Ready** | OFF — Milling motor Switch on |
| Milling motor Switch on | x | x | x | x | = shall? | x | 1 | Milling motor = on | Move the milling machine into position ready | **Set** milling motor to Target speed | Ready |
| Ready | = *ready* | x | x | 0 | x | Directional drive — x | 1 | x | Move milling machine to *rest* position — x | Command **OFF** — Command **Save Energy** — Command **Milling** | OFF — Save energy — Milling active |
| Milling active | = *work* | Synchronous position | = 1? | 0 | = shall | Directional drive | 0 | For masterpos. *Trigger* + *x* Move milling machine to position *work* Suction = on | For masterpos. *Trigger* + *y* Move milling machine to position *ready* Suction = off | Milling machine in position *ready* | Ready |
| Malfunction | x | x | x | x | 0 | 0 | 0 | Milling motor = off Infeed motor = off | x | Command **Quit** | Reference-ride |

### 2.4.1.3  *The interaction diagram*

The representation of the Moore automaton in the selected description form already shows in detail the individual interactions occurring via inputs and outputs or service calls (referred to as *commands in the table and state diagram*) which are required by the individual functions. To design an interaction diagram, the individual service calls must now be extracted, and the required service input and output data defined.

For the *back processing* module, Table 2.4 shows excerpts of the service calls for the external functions and the corresponding internal

**Table 2.4.**   Overview of service calls of the back milling module.

| Service call | Service input data | Functionality | Service output data |
|---|---|---|---|
| Module Switch on | None | Performs a system test and moves the milling head to the rest position. | *OK* or *Errorcode* |
| Module Switch off | None | Moves the milling head to the park position. | *OK* or *Errorcode* |
| Module in energy-saving mode | None | Moves the milling head to the rest position and switches off the milling motor. | *OK* or *Errorcode* |
| Module in standby mode | None | Switches the milling motor on and moves. The milling head to the standby position. | *OK* or *Errorcode* |
| Milling | Technological target parameters $a_1$, $a_2$, $z$ | Executes the milling process synchronized to the master position with the technological target parameters when a product arrives. The milling head is then moved back to the standby position. | *OK + actual parameters* or *Errorcode* |
| Acknowledge fault | User-ID | Change to state *Reference run* and wait for the state *Ready* to be reached. | *OK* or *Errorcode* |

functions. In the actual implementation, this overview still needs to be supplemented by labeling the service types and identifiers.

Table 2.5 contains an example of the service call with which a perfect binder could be commissioned to process an individual product. Additionally, to the ID of the specific product, the service input data would also contain the identifiers of the associated materials (book block, binding, adhesive type) as well as the production parameters.

**Table 2.5.** Excerpt from the service call for the execution of a perfect binding.

| Service call | Service input data | Functionality | Service output data |
|---|---|---|---|
| Bind product | Product ID Book block ID Cover ID Adhesive ID Technological parameters | Service calls to modules: 1. ... 2. Back processing *Milling the back* 3. Adhesive station *Applying glue* 4. ... | *OK + actual parameters* or *Error code* |

The functionality of this service call can be described in the *interaction diagram* of the perfect binder as follows. It is again assumed that it is integrated into the digital production environment as an I4.0 component and is managed as an entity (CP class 44).

The simplified illustration shows the administration shell of the perfect binder with the component manager in the form of a *scheduler* that controls the technological sequence of the process based on the information stored in the manifest. The process begins with a service call to manufacture a product with individual parameters. The first step is to check whether the machine is technically and organizationally ready. This includes determining whether all required production steps can be carried out in accordance with the product specifications. This would be, for example, checking whether the appropriate hotmelt adhesive derivative is available in the system. If this is not the case, the order is rejected with a corresponding status message.[45] When the order is accepted, all specified

---

[45] In this case, a decision must be made at the production management level as to whether additional perfect binders should be requested or whether a service call should be sent to the service personnel for conversion.

process steps are activated in sequence by service calls and confirmed with a status message once they have been completed.

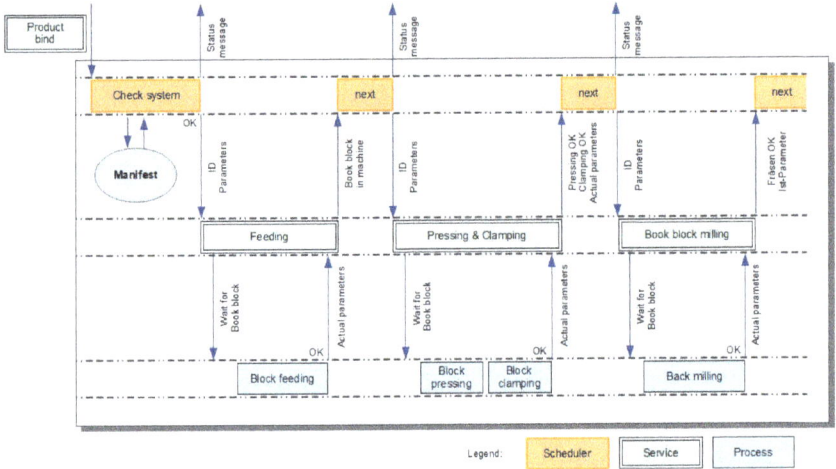

**Figure 2.31.**    Section of an interaction diagram for the adhesive bonding process.

Whether these messages, as depicted here, are transmitted to the production control system is a design decision in the specific implementation. What is not shown here is the complete error handling, for which separate representations of the interaction diagram are practicable in the interests of better clarity (Figure 2.31).

## 2.4.2  Design of a modular construction

The design aspects must now be incorporated into the *functional structure* model during the design process. For this purpose, the functional elements are considered in their technological sequence and are correspondingly constructively underpinned. The result is the concrete model of an individual production system, which is formed from the set of basic modules and their derivatives.

2.4.2.1  *The module diagram*

While the aforementioned diagrams should primarily be used as a precursor for software engineering, the *module diagram underlying* the approach in this book serves more constructive aspects. In this step, we want to design the actual modular structure and assign each technological function to specific mechatronic units. The character of the individual modules must also be determined, i.e. whether, for example, a function is to be implemented as a mechatronic unit alone or can be part of a more complex module. The question of whether the modules should be integrated, autonomous, or modular themselves must also be clarified. Finally, all units must be brought together with the components of the *backbone*.

The starting point of these considerations is now not only the individual basic functions but also the individual results of the target analysis according to Figures 2.16–2.21. Following are some selected aspects that are considered for our example of *perfect binders* and will be examined in more detail in the further course.

**Product: Volatility and quantity**
The results of this analysis lead to the definition of a minimum and an average batch size for the products. As we are working in the context of Industry 4.0, let's assume that we want to manufacture individual products in batch size 1, but how far does this specification go? Does it really mean that each individual product has different formats, binding processes, and covers, or is the production control system able to create larger batches with the same format and identical binding process? In the latter case, individual products are also created which differ solely in terms of content and the number of pages. As a result, however, the changeover effort is lower. An automated format changeover could also result in the quantitative product throughput being correspondingly high.

However, if the volatility and required quantity are that high assuming that it will not be possible to produce reasonably large batch sizes, it would be counterproductive if the *gluing station* had to be constantly replaced by an operator, for example. In this case, one could consider the parallel integration of the various derivatives of all relevant functional units and the development of an intelligent material flow system within the machine.

**Product: Quality and quantity**
For our perfect binder example, we want to assume that the dwell time after *joining* and *pressing* is decisive for the durability of the final product. For example, if the bound book is removed from the pressing device too early, the adhesive may not have hardened enough leading to the first cracks and detachments during subsequent handling. Therefore, the productivity of the perfect binder is directly related to the quality, as a longer dwell time means a reduction in performance without appropriate design measures.

**Investment and environment: Availability in digital production**
High product volatility and the digital environment also require maximum availability of the production system. Costly changeovers should therefore be avoided and, if necessary, cushioned by intelligent setup scenarios.

**Company: Production of perfect binders according to individual customer requirements**
A broad portfolio of different types requires efficient production of the machines.

This selection of criteria is now be applied to two perfect binders for the production of individual photo books.

**Example 1: Individual perfect binder for high productivity with lower product volatility**

- Larger batch sizes with the same binding technology can be created by combining products of the same type.
- Only books of the same type with soft covers but different adhesives should be produced.
- Only selected and predefined formats are possible.
- The focus is on maximum productivity and product quality.
- The integration in a digital production requires CP-Class 44.

- The development of individual machines with short delivery times and a low investment budget requires efficient production.

**Example 2: Individual perfect binder for high productivity and very high product volatility**

- Batch size 1 should be possible for all formats and binding processes.
- Books can be produced with different cover variants.
- Any number of freely defined formats is possible within technological limits.
- The focus is on maximum product volatility and quality.
- The integration in a digital production requires CP-Class 44.
- The development of individual machines with short delivery times requires efficient production.

The examples have been deliberately chosen so that there are overlaps in the requirements.[46] These include the environment of digital production, high productivity, and the individual configuration of machines with short delivery times.[47]

So far, we have shown that the latter in particular generates the requirement for a modular design. At the beginning of the considerations, the technological functions must therefore be assigned to individual modules. The ERDs of the individual functions serve as the basis for this (see example of back milling, Figure 2.28). A decision must then be made as to which module versions are practicable for the individual submodules and how constructive compatibility with each other and with the backbone should be ensured. Table 2.6 shows an example of this for the *book block processing* area of a perfect binder.

---

[46]It is common practice for a company to have to process production orders with different and overlapping requirements at the same time.

[47]This fact can also be a requirement for the manufacturing company in order to create a unique selling point on the market.

**Table 2.6.** Overview of the module classification for a perfect binder, book block processing area.

| Module | Module type | Functionality | No. type | Assembly | Material flow | Interaction features | | | | | |
|---|---|---|---|---|---|---|---|---|---|---|---|
| | | | | | | Electric | Suction air | Blown air | Safety | HMI | 140 |
| Block feed | Integrated | Feeding | 1 | Fixed | Synchronous mechanical | 80 VDC 10 A / 24 VDC 3,5 A | 0.3 l/product | 0.2 l/product | Safe Torque Off | x | CP12 |
| Jog station | Integrated | Jogging | 1 | Fixed | x | 400 VAC 16 A | x | x | Safe Torque Off | x | CP12 |
| Press station | Integrated | Pressing clamping | 1 | Fixed | Asynchronous electric | 80 VDC 10 A / 24 VDC 0,8 A | x | x | Safe Torque Off | Keys | CP12 |
| Back mill | Integrated | Back milling | 2 | Latching | Synchronous electric | 400 VAC 16 A / 80 VDC 10 A / 24 VDC 3,5 A | 2.5 l/product | x | Safe Motion | Mini-display | CP23 |
| Glue station | Autonomous modular | Apply adhesive | 3 | Latching | Synchronous electric | 400 VAC 16 A / 24 VDC 2,5 A | x | 0.2 l/product | Safe Motion | 5.7" Touch | CP44 |
| Stub station | Autonomous | Stub apply | 1 | Latching | Synchronous electric | 80 VDC 10 A / 24 VDC 2,0 A | 0.3 l/product | 0.5 l/product | Safe Motion | Mini-display | CP23 |

Book block processing

The entries have the following meanings:

### Module definition
Module types are defined, and their basic functionality is assigned.

### Number of types (derivatives)
It is noted whether any derivatives of the module exist.

The *number one* means that there is exactly one version of this module, and it can therefore only be used in unchanged form. A *larger number* indicates that different derivatives of this module are required. However, this does not mean that all derivatives must occur in one machine (see cardinalities in the ERD, Figure 2.4).

### Assembly
In principle, this defines how the module is integrated into the production system.

*Fixed* means that it is permanently mounted and no replacement in the sense of a technology-dependent retrofit is planned. This assembly can be detachable (e.g. as a screw connection) or non-detachable (e.g. as a welded connection).

A *latching* assembly indicates that a mobile module can be easily installed and removed by operating or service personnel, as may be necessary for retrofitting, for example.

### Material flow
The type of access to the material passing through is defined.

*Mechanical* means that the module is connected to the transport system (e.g. a collecting chain) via a rigid output (e.g. a mechanical gearbox). The coordination is usually synchronized.

If it is an electric gearbox with servo drive technology, *synchronous* coordination will generally be required. *Asynchronous* means, for example, that only speed synchronization is required instead of position synchronization. This statement has a considerable influence on the constructive and dynamic design of the automation system. At this point, it would also be possible to note the dynamic requirements such as speed and position accuracy.

### Media supply
The characteristic values are specified for all *supply media*.

### Safety
The *safety requirements* of the individual modules must be specified for the design of the automation system. In the case of an emergency stop,

this can be a simple shutdown of the drives (Safe Torque Off) or more complex safe drive functions such as Safe Limited Speed.

**Operating and monitoring (HMI)**

If separate local components are required to *operate* the individual modules, this also has a significant influence on the design of the automation system. It is therefore important to consider whether operation should only be carried out via individual buttons or by means of a graphical user interface, for example.

## I4.0 properties

If the production system is operated in a digital production, it is necessary at this point to specify whether and how the respective module is identified and communicated. Using the glue station as an example, we assume that it is to be kept ready by preheating at an external docking station. It is important for the production control to know whether there is a ready-to-operate glue station or whether it needs to be preheated further. Therefore, this module is classified as I4.0-compliant and as an autonomous module with *CP-class* 34.

The module derivatives must now be specified in a further representation. In Table 2.7, this is shown for the back router and gluing station modules.

The entries have the following meanings:

**Back processing**

The variants differ in the possibilities of the milling process. With type 1, only the entire spine can be milled. This means that *flex-stable binding* is not technologically possible, and the technical design can be made much simpler and therefore cheaper by eliminating the position-dependent infeed of the milling head. Type 2, on the other hand, is suitable for both binding processes, as full milling is possible with the parameters $a_1 = a_2 = 0$ and any other milling is possible with $a_1$, $a_2 > 0$.

There is yet another issue. In Section 2.1.4 we learned that modules can only be replaced without restriction if they are designed to be functionally and connection-compatible. The *functional compatibility* of back processing is demonstrated by the fact that both derivatives can execute the *back processing* service. The only difference is that with type 1, the service call is only accepted if the parameters $a_1 = a_2 = 0$. The module types are connection-compatible due to the same interactions, even if type 1 only requires a subset of type 2, because it is irrelevant for the compatibility whether and how an offered service is used or not.

**Table 2.7.** Classification and properties of some module derivatives for a perfect binder.

| Module | Module type | No. type | Description | Functionality | Assembly | Interaction features | | | | | | |
|---|---|---|---|---|---|---|---|---|---|---|---|---|
| | | | | | | Material flow | Electric | | Suction air | Blown air | Safety | HMI | 140 |
| Back processing | Integrated | 1 | Only complete milling of the book block possible | Full milling | Mobile resting | x | 400 VAC 24 VDC | 16 A 1,5 A | 2.5 1/product | x | Safe motion | Adjusting screws | CP23 |
| | | 2 | Complete and partial milling of the book block possible | Full milling partial milling | Mobile resting | Synchronous electric | 400 VAC 80 VDC 24 VDC | 17 A 10 A 3,5 A | 2.5 1/product | x | Safe motion | Mini-display | CP23 |
| Glue station | Autonomous modular | 1 | Back and side gluing of the book block with hot-melt adhesive | Adhesive apply | Mobile resting | Synchronous electric | 400 VAC 24 VDC | 16 A 2,5 A | x | 0.2 1/product | Safe motion | 5.7" Touch | CP34 |
| | | 2 | Back and side gluing of the book block with polyurethane adhesive | | Mobile resting | Synchronous electric | 400 VAC 24 VDC | 16 A 2,5 A | x | 0.2 1/product | Safe motion | 5.7" Touch | CP34 |
| | | 3 | Back and side gluing of the book block with dispersion adhesive | | Mobile resting | Synchronous electric | 400 VAC 24 VDC | 16 A 2,5 A | x | 0.2 1/product | Safe motion | 5.7" Touch | CP34 |

Derivatives

**Gluing station**

This module neither has functional nor design differences on the outside. Only the internal structure and the internal functions for processing and dosing the various types of adhesives are different.

### 2.4.2.2 *Qualitative module scheme*

With the information obtained so far, it is now possible to design a qualitative module scheme. For this purpose, the positions of the individual modules in relation to each other and in relation to the backbone and transport system are outlined. Figure 2.32 shows a module diagram for case study 1.

In the example shown, the material transport system could be designed as a circulating transport chain to which all modules are connected according to their specifications. The backbone components are located within the machine body. It should be possible to replace the *gluing station* and *cover processing* modules with derivatives for different adhesives or covers in case of a change in technology. Although two derivatives are available for back processing, only type 1 is used for full milling in this configuration and is therefore assigned the fixed attribute (shown in light color in the image), unlike the gluing station and cover processing.

This diagram clearly shows the advantages and disadvantages of a continuous collection chain as a transport system. The basic advantage of this configuration is certainly the price of the transport system. A continuous collecting chain with a main drive and associated servo technology is well established in terms of design and can be manufactured with comparatively little effort. However, the decisive disadvantage of this concept

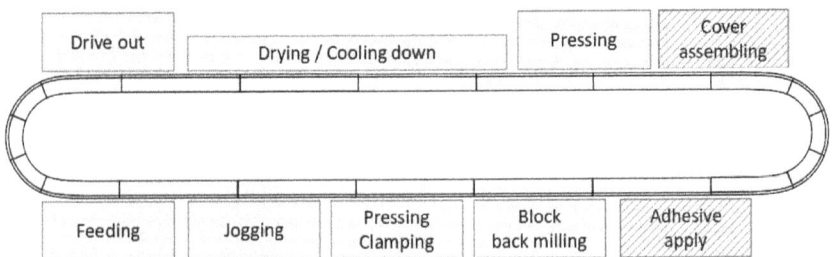

**Figure 2.32.**   Qualitative module scheme for a perfect binder with a collecting chain as a transport system.

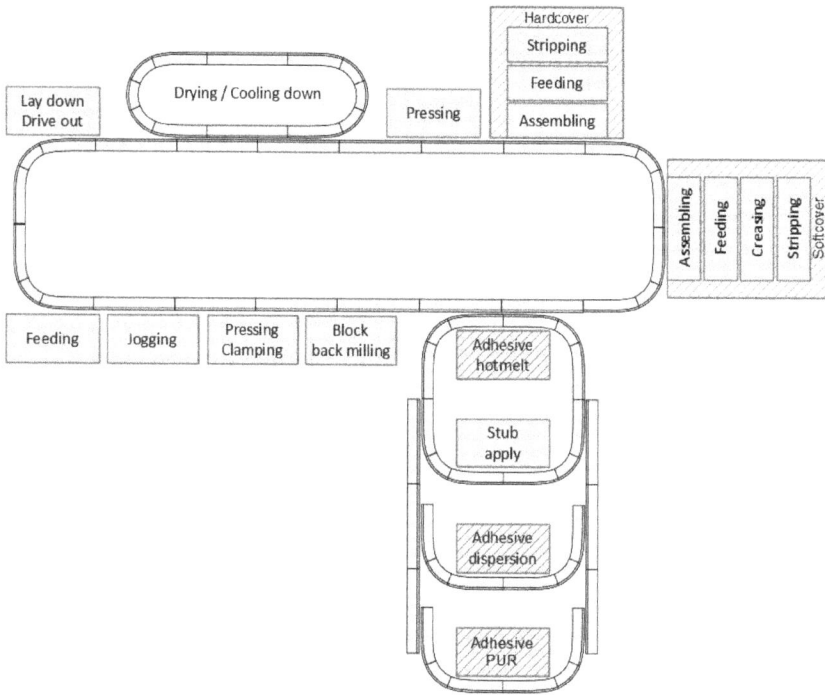

**Figure 2.33.**   Module scheme for a perfect binder according to case study 2 with a long-stator linear motor as transport system.

is productivity, as the product that requires the longest dwell time determines the productivity of the overall system.

The module scheme for case study 2 is demonstrated in Figure 2.33. This configuration is based on a transport system as presented in [25] and fulfills the requirement for the greatest possible flexibility for products in batch size 1. Each product can differ in this system according to technological limits, both in the formats and in the binding processes.

To ensure this, the following details have been considered:

- The transport system is designed as a *long-stator linear motor* and equipped with electric switches that can change direction within a few milliseconds. For product tracking, the technology data of the individual product can be read directly before each branch via an RFID reader or a QR code scanner. For example, if after back processing the service call "Apply adhesive/hotmelt" is recognized as the next technological step, the central backbone controller can generate a corresponding service as a new transport order for precisely this shuttle, using the data stored in the manifest.[48]

- In the interests of efficiency, the individual modules are essentially arranged on the transport system according to the technologically required sequence. However, this is not necessary, as illustrated by the example of the stub feeder.

- Because a hot melt adhesive is preferably used for the assembly of a stub, this is located directly downstream of the corresponding gluing station. As shown in Figure 2.33 the product passes this gluing station again directly after the stub feeder and receives the second adhesive application for the binding assembly. If necessary, the shuttle can also remain on the secondary line to allow the glue to cool down a little or it can pass the gluing station without being processed if, for example, no binding is to be fed for a product. Or the module for stub assembly is installed further away (e.g. at the location of the PUR gluing station). In this case, it would be fed at a high speed after the first adhesive application and integrated back into the product flow at a slower rate. This example shows that the possibility of using a module several times means that the system can also be set up in a more space-saving and cost-effective manner with maximum flexibility.

- When a product reaches a module, the technology data can be read out directly and processing then takes place according to the specified data and/or the data updated by the previous processing steps. For example, the pressing module can measure the actual thickness of the book block as the product passes through and then write it to the RFID chip as an actual value. With this information, all further processing steps that require the exact block thickness in their technology can be carried out at higher quality.

- With the technology presented in [25], the drive functions of the modules can be synchronized directly and with high accuracy to the position of the shuttles. This makes it possible, for example, to adjust the transport speed of the book block around a module depending on the technology without any repercussions for the other shuttles. Position-synchronized spine processing according to Figure 2.29 is also possible with this system.

---

[48]The system presented in [25] also has the option of assigning the technology data of a product directly to a shuttle. This means that the processing steps can be transferred to the transport system at the time of import or by the ERP, and there is no need for direct queries before the processing stations.

- In order to meet the demand for high quantity and quality at the same time, all possible module derivatives are integrated into the production system at the same time but can be easily exchanged, rearranged, or added to due to their *mobile latching* design. This minimizes downtimes and changeover times. Furthermore, parallel lines can be installed to ensure the required dwell times, which can also be run through several times depending on the technology used. This means that the total throughput time of a product does not depend on products with longer dwell times.
- The transport system is also modular and can be designed in a way that reserves spaces for additional modules can be provided or retrofitted. For example, the system can be equipped at a suitable position with a module for additional printing with a QR code as a shipping label or for inserting a reading tape, without affecting the operation of the other modules.

This solution approach demonstrates how innovative technologies enable completely new possibilities, especially for the manufacture of individual products, and also generate unique selling points. The flexibility and quantity that can be achieved would certainly not be possible with a circulating transport chain. In addition, there is the option of extending the transport system across the entire book production line from digital printing to block making, perfect binding, trimming, and packaging. On the one hand, this results in synergies due to the elimination of infeed and outfeed modules in the individual machines, and on the other hand, the product flow can be designed to be highly effective and I4.0-compliant. This is the case, for example, in a modular beverage bottling line where each bottle can be individually filled and printed [26]. The advantage of the object and service-oriented concept is also evident, as none of the modules necessarily needs to know anything about the overall system and works completely independently.

### 2.4.3 Design of the automation system

The module diagram now serves as a template for the further mechanical, electrical, and automation engineering process.[49]

---

[49] In order to do justice to the purpose of this book, only the automation processes are considered in the following.

### 2.4.3.1 *Hardware concept*

For the development of the hardware concept, the respective module type must first be considered (Tables 2.6 and 2.7). While an integrable module does not need to have its own control intelligence, this is a mandatory requirement for autonomous modules. The information in the *Safety* and *HMI columns* considers the requirements determined for this. Figures 2.34–2.36 show examples of possible automation concepts for the three module types, although these are only rough illustrations.

For case study 2, the basic automation concept is shown in Figure 2.37.

It depicts the modular arrangement on a common field bus, which is responsible for transmitting the asynchronous services for technology, operation, and service but also has to transport synchronous data in real time. This results in several design requirements with regard to openness, real time, and safety behavior, which are discussed in more detail in Section 4.4.

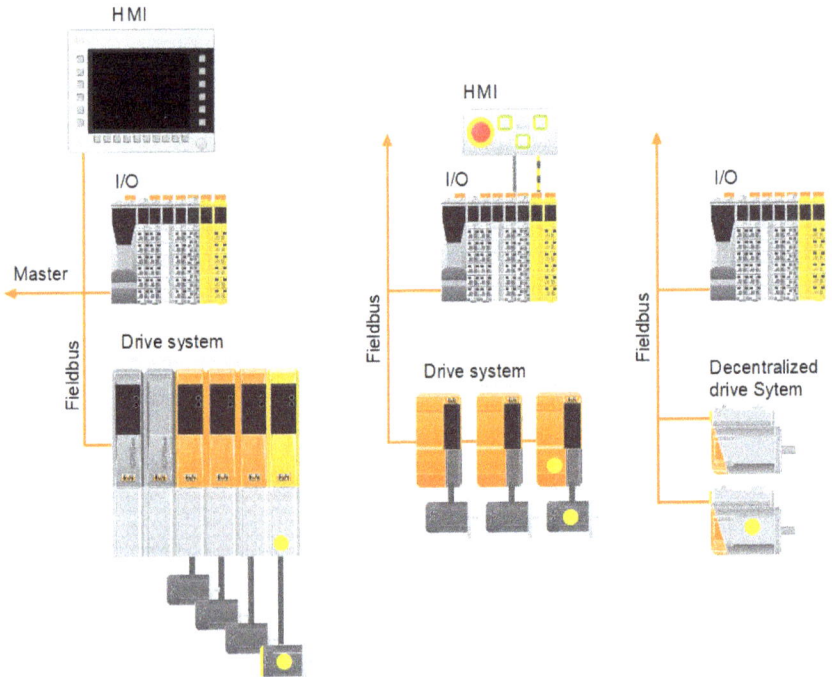

**Figure 2.34.**    Structural examples of automation for integrable modules.

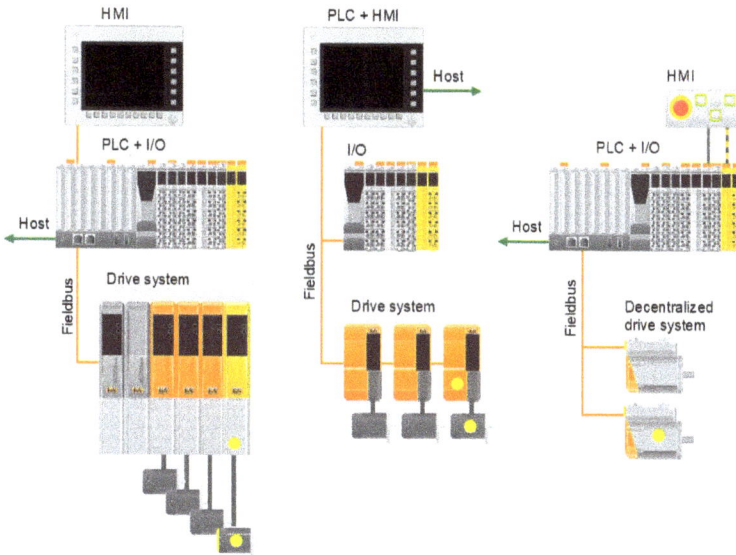

**Figure 2.35.** Structural examples of automation for autonomous modules.

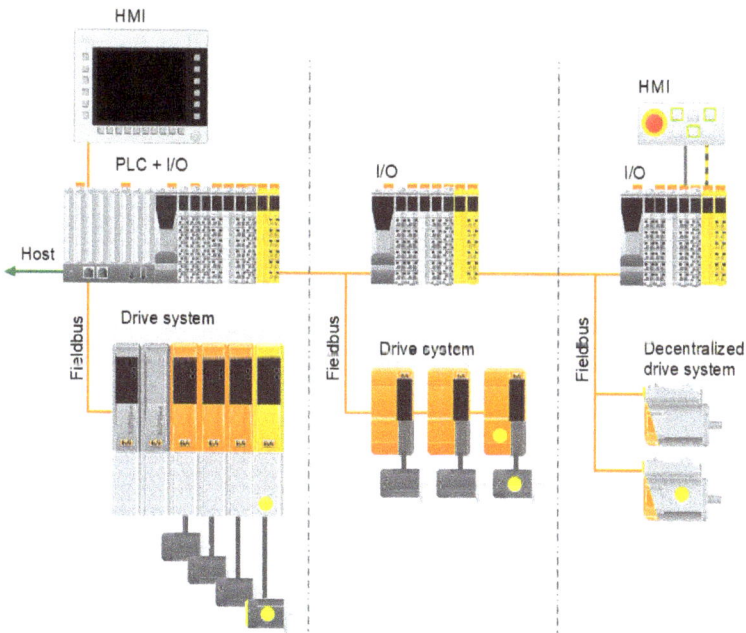

**Figure 2.36.** Structural example of automation for a modular module.

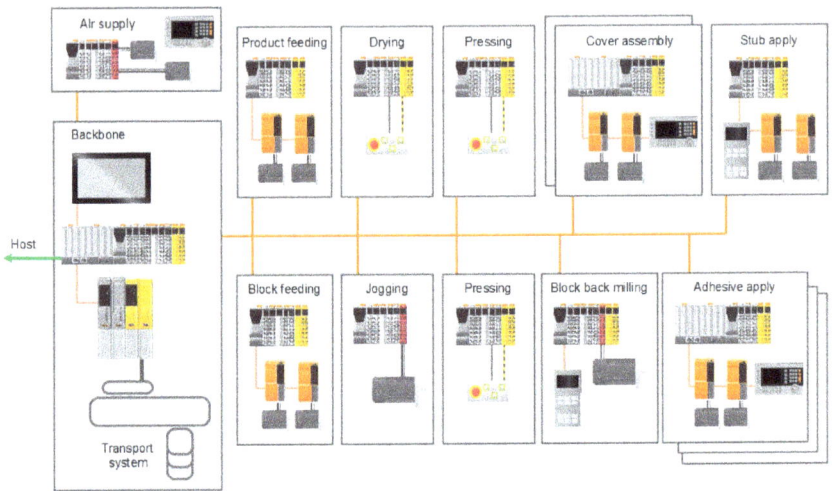

**Figure 2.37.** Automation concept for a perfect binder according to case study 2.

At this point, however, it should be mentioned that the consideration of synchronous data traffic is part of the module automation design decision from the outset. If, for example, it is defined that the products are always to be detected when entering the working area of a module, e.g. via a print mark recognition or a trigger signal (Figure 2.9), the transfer of the synchronous master position can possibly be designed to be more resource efficient. This provides additional scope for a dynamically feedback-free system design, especially when the number of modules is high and fluctuates greatly.

The considerations regarding machine safety and the integration of safety components are even more in-depth, as a modular structure has additional requirements in this respect. These are described in detail in [27] and are dealt with in more detail in Section 4.3.

### 2.4.3.2 *Software concept*

The software concept can now be designed from the results of the previous engineering process, as shown in the previous sections. The design process for a modular and object-oriented software structure begins with the respective ERD and the assigned properties. However, there are some aspects of modular engineering that are particularly important to consider during software development and which will be discussed in more detail here. Firstly, these are the aspects that result from the typing of the modules.[50]

## Autonomous modules

This type of module has its own automation system with the necessary compatibility for the exchange of all operational data for the technological functions, such as position data for synchronizing movements, but also for machine safety, diagnostic information (condition monitoring), or operation (HMI). This is usually done via a field bus, which also meets the requirements for synchronous data traffic with its real-time behavior. For a service-oriented, real-time capable, and I4.0-compliant data exchange, Ethernet has already been defined as the physical medium and OPC UA TSN as the protocol format as a standard recommendation.[51]

Because of the autonomy of this type of module, the module's own automation technology can be designed with complete freedom if these requirements are considered. The module software is created with the required tools and is not bound to specific conventions such as specific manufacturers, programming languages, or operating systems if the compatibility criteria are met. However, this situation poses both an opportunity and a challenge, as the behavior of the module as a black box can also mean that the engineering team of the machine or system manufacturer only influences the design of the automation technology through a precise and conscientious specification. On the other hand, hardware and software that is precisely tailored to the technological task can be used to generate technical and/or economic benefits. The selection of a suitable automation system is therefore part of the design decision for each module from the very beginning.

## Integrable modules

In terms of automation, this type of module is operated by the backbone controller, as integrable modules do not contain any technological intelligence of their own.[52] While the creation of modules at the hardware level is quite straightforward through the use of fieldbus nodes, there is a risk

---

[50] See also Section 2.1.1.

[51] See status report on the development of an interaction model [11].

[52] The term "no technological intelligence" is intended to make it clear that individual control components such as servo drives or I/O modules can certainly be intelligent and

in software design that any modularization efforts made beforehand may be undone by an unsound separation of the individual program components. In this respect, the use of function blocks, macro structures, or multitasking operating systems can only be a first step toward a clean separation of clearly defined software modules. This conflict is exacerbated by the fact that software is sometimes developed by multiple project teams operating remotely around the world. Therefore, an engineering system that only establishes the connection between logic and physics via a configuration layer is a basic requirement for this module variant.

**Modular modules**
The aforementioned aspects also apply to this type of module. The only difference is that the horizon of consideration is extended to the overall module via the sub-modules. A modular software structure is often dispensed with too quickly simply because of the smaller scope of the functions to be realized. The same rules of modularization also apply to this type of module with at least the same synergy effects.

With these aspects in mind, the software development tool in particular must be carefully selected for the individual module types. The tool shown in Figure 2.38 demonstrates how a modular software structure can be realized.[53]

The functions and modules of the backbone as well as the integrable modules such as *block feeder* and *jogging station* are managed as separate application modules. These contain the program code for the technological functionalities, drive, and safety technology through operation and visualization (Figure 2.39). The modules can be developed separately by separate teams at different locations and installed on the PLC parallel to other modules. It is also evident that the autonomous *air supply* and *glue station* modules only exist with one communication object. The same applies to the I4.0 administration shell and the required manifest as a data object.

This results in a clear separation of physics and logic. This in turn supports the developers in the creation of derivatives, as the software

---

equipped with processors but are always dependent on a central control system via which the technological process is controlled.
[53] Illustrations from the B&R engineering system *Automation Studio* [36].

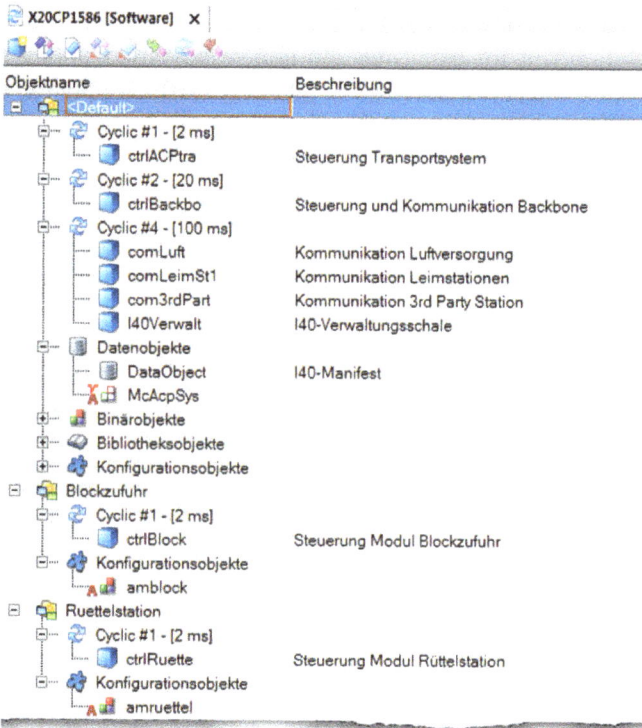

**Figure 2.38.** Illustration of the application modules of a perfect binder using the example of B&R Automation Studio (excerpt).

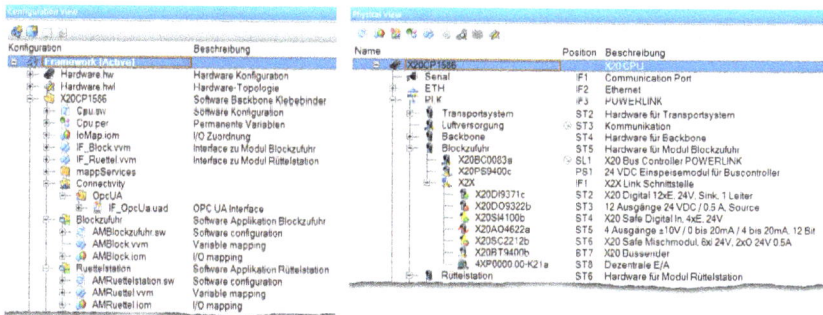

**Figure 2.39.** Assignment of logic to physics via a configuration layer using the example of B&R Automation Studio for a perfect binder (excerpt).

components can be developed independently of their physical assignment. The requirement for extensive and non-reactive decoupling of the individual sub-projects is therefore also largely fulfilled in the software.

## 2.5  Summary and conclusion

In this chapter, we have established how a modular production system can be designed. Based on the fundamentals of modularization, we have defined module types, described their properties and connections through object-oriented programming methods, and used examples to show how they can be applied to mechatronic systems. We also described how a production system can be integrated into the digital production environment and how I4.0-compliant design must be incorporated into the design process from the outset. To this end, we analyzed the current state of development from the status reports of the Industry 4.0 platform, presented the aspects relevant to mechanical and plant engineering in broad outline, and demonstrated them using examples. Finally, the design process, from detailed analysis through the decomposition of functional expertise to constructive and automation-related conceptualization, was applied using a specific machine as an example. For a better understanding, we will present a successful example from the textile processing industry, in which the essential elements of this design process were applied. These are plants for fiber preparation in which raw material (e.g. cotton or man-made fibers) is prepared in a multi-stage process [28]. The technological process begins with the opening of the raw material bales in the so-called *blowroom* (Figure 2.40).

In the next step, the raw material is cleaned in several stages. This prevents plant residues, insects, dust or other foreign substances such as plastic residues or metal shavings from entering the final product. In addition, fiber blends are already produced in this process phase, in which recyclable sliver waste and/or various raw materials are added to the product stream. The fibers are then parallelized by carding and formed into either nonwoven or fleece[54] (Figure 2.41).

The challenge with these systems is that different process steps are required depending on the raw material to be used and the end product to

---

[54] Sliver refers to the roving yarn that is subsequently spun into threads in the spinning mill. In highquality slivers, the fibers run parallel and very evenly.

**Figure 2.40.** Bale opener for removing the raw fibers.
*Source*: Trützschler GmbH & Co. KG.

**Figure 2.41.** Finished sliver for the subsequent spinning process.
*Source*: Trützschler GmbH & Co. KG.

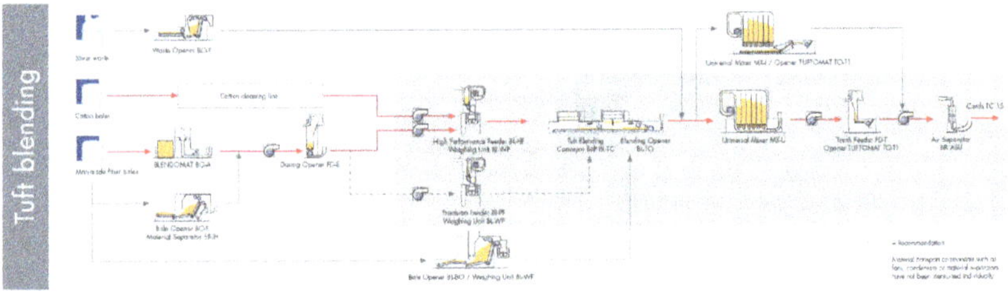

**Figure 2.42.**    Block diagram of a blowroom for fiber preparation.

*Source*: Trützschler GmbH & Co. KG.

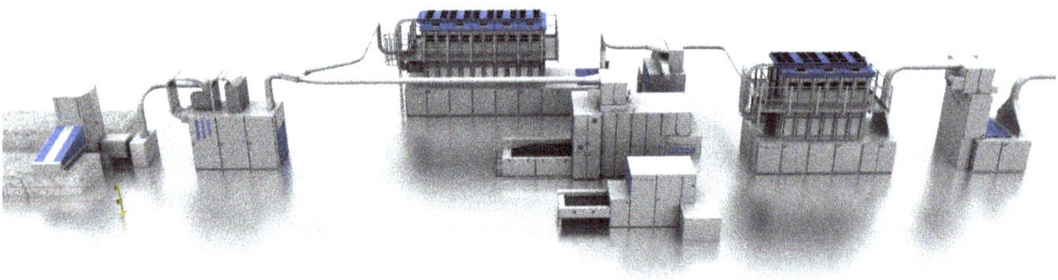

**Figure 2.43.**    Plant diagram of a blowroom for fiber preparation.

*Source*: Trützschler GmbH & Co. KG.

be manufactured. In addition, the fibers are subjected to a great deal of stress during processing. Therefore, the number of necessary process steps must be kept as low as possible. To achieve this, precise coordination between the individual machines is crucial, as the choice of the right cleaner or a combination of several depends on the raw material and the production output. These requirements mean that the entire system must be individually planned and equipped with the necessary machines. In addition, the customer must be able to individually determine the inter-connection of the individual machines from order to order. Figures 2.42 and 2.43 illustrate a typical example of a plant in which a high-quality fiber blend can be produced from cotton, man-made fibers, and reusable production waste.

To achieve this, the system manufacturer has strictly modularized the system structure and defined each individual unit and each section of the transport route as independent objects. These entities are not only assigned the technological functions and properties, but also all the data required for order-specific production, operation, and service. This enables extremely effective processing of all internal processes, from planning and production through to delivery and commissioning. According to the development manager, the workload for this process has been reduced from 4 man-weeks to 16 man-hours. In addition, the end-to-end process results in significantly fewer complaints and rework.

In terms of automation technology, the individual machines are designed as autonomous or autonomous modular modules. The pipe segments not equipped with control technology exist as virtual objects. As T-pieces act as switches and therefore have sensors and actuators, they are also managed by a machine controller as integrable modules, whereby an autonomous approach is also conceivable for these units.

A key paradigm of this object-oriented approach is that the individual modules only interact with their immediate neighbors. They only receive information via the product flow and the information flow in the opposite direction. This means that there is no need for central intelligence for control and regulation tasks, as the system configuration is always detected via the objects and their communication with each other. Instead, the tasks are organized within the objects and between the units themselves. For the actual control and regulation tasks, none of these objects need to know the entire system — knowledge of the immediate neighbors and communication with them is sufficient. A higher-level instance is only necessary because there must be an interface in the plant via which recipes and orders enter the system and via which the process can be configured and visualized.

The functionality during operation is almost simple and extremely logical. Each unit analyzes the incoming product according to the criteria that are relevant to the respective technological sub-process. If, for example, it is determined that the incoming flow rate is not sufficient to achieve the required output, a corresponding message is sent to the upstream station. If this station is unable to meet the requirement, the information is passed on and the operator is only requested to intervene at the end. For example, by activating an additional bale opener or by reducing the required system output. This method of operation enables the operator to flexibly configure the system and is easy to expand and modify if requirements change. The machine modules themselves go through a completely decoupled life cycle so that the manufacturer's innovations can be introduced to the market without affecting the rest of the product portfolio.

# References

[1] Wildemann, H.: Modularization in Organization, Products, Production and Service (TCW-Report, No. 66), Munich 2014.

[2] Fuchs, J., Legat, C., Kernschmidt, K., Frank, T. and Vogel-Heuser, B.: Interdisziplinärer Produktansatz zur Unterstützung der Wiederverwendbarkeit im Maschinen-und Anlagenbau, in: *Tagungsband EKA,* Magdeburg 2014.

[3] Krusche, T., Leyers, J., Oehmke, T. and Parr, T.: Bewertung von Modularisierungsstrategien für unterschiedliche Fahrzeugkonzepte am Beispiel des Vorderwagens, in: *ATZelektronik* Nr. 10/2004 (106).

[4] ITQ GmbH: Kompetenz in Mechatronik Software und Systemen, Garching 2017, available online at: https://cms.itq.de/wp-content/uploads/2016/10/ITQ_Unternehmensbroschuere.pdf, last accessed: 01.07.2017.

[5] NAMUR — Interessengemeinschaft Automatisierungstechnik der Prozessindustry e. V. (ed.): Anforderungen an die Automatisierungstechnik durch die Modularisierung verfahrenstechnischer Anlagen (NAMUR Recommendation NE 148), Leverkusen 2013.

[6] Meyers Lexikonredaktion (ed.): Art. Kompatibilität, in: Meyers großes Taschenlexikon in 24 Bänden, Vol. 12, 4th revised edition, Mannheim, Leipzig, Vienna, Zurich 1992.

[7] Thielicke, R.: Industry 4.0. Fraunhofer-Institut arbeitet am USB-Prinzip für die Fertigung, in: *Technology Review* v. 04.12.2013, available online at: http://www.heise.de/-2059877, last accessed: 24.06.2017.

[8] Verein Deutscher Ingenieure e. V. VDI/VDE Gesellschaft Mess-und Automatisierungstechnik (Hrsg.): Auf dem Weg zu einem Referenzmodell (Industry 4.0 — Statusreport), April 2014.

[9] Plattform Industry 4.0: Referenzarchitekturmodell Industry 4.0 (RAMI 4.0), in: Plattform Industry 4.0, available online at: http://www.plattform-i40.de/I40/Redaktion/DE/Downloads/Publikation/rami40-eine-einfuehrung.html, last accessed: 04.08.2017.

[10] Plattform Industry 4.0: Fortentwicklung des Referenzmodells für die Industry 4.0-Komponente, in: Plattform Industry 4.0, available online at: http://www.plattform-i40.de/I40/Redaktion/DE/Downloads/Publikation/struktur-der-verwaltungsschale.html, last accessed: 04.08.2017.

[11] Plattform Industry 4.0: Weiterentwicklung des Interaktionsmodells für Industry 4.0-Komponenten, in: Plattform Industry 4.0, available online at: https://www.plattform-i40.de/I40/Redaktion/DE/Downloads/

Publikation/interaktionsmodell-i40-komponenten-it-gipfel.pdf?__blob= publicationFile&v=12, last accessed: 04.08.2017.

[12] Plattform Industry 4.0: Example Use Case Definition, Models and Implementation (Workingpaper 2017), in: Plattform Industry 4.0, available online at: https://www.plattform-i40.de/I40/Redaktion/DE/Downloads/ Publikation/Industry-40-%20Plug-and-Produce.html, last accessed: 04.11.2017.

[13] Horn, C., Kerner, I. O. and Forbrig, P.: Lehr und Übungsbuch der Informatik, Bd. 1: Grundlagen und Überblick, 2nd ed., Leipzig, Munich, Vienna 2001.

[14] Horn, C. and Kerner, I. O.: Lehr und Übungsbuch der Informatik, Bd. 3: Praktische Informatik, Leipzig, München, Wien 2001.

[15] Grützner, J., Höme, S. and Diedrich, C.: Semantic Industry: Herausforderungen auf dem Weg zur rechnergestützten Informationsverarbeitung der Industry 4.0, in: *Tagungsband EKA*, Magdeburg 2014.

[16] German Electrical and Electronic Manufacturers' Association (ZVEI) e. V.: The Reference Architecture Model Industry 4.0 (RAMI 4.0). An introduction, in: Plattform Industry 4.0.

[17] Plattform Industry 4.0: Forschungsagenda Industry 4.0 — Aktualisierung des Forschungsbedarfs, in: Plattform Industry 4.0, available online at: http://www.plattform-i40.de/I40/Redaktion/DE/Downloads/Publikation/ forschungsagenda-i40.pdf?__blob=publicationFile&v=5, last accessed: 04.08.2017.

[18] Verein Deutscher Ingenieure e. V. VDI/VDE Gesellschaft Mess- und Automatisierungstechnik (Hrsg.): Industry 4.0 – Technical Assets Basic terminology concepts, life cycles and administration models, March 2016, in: Plattform Industry 4.0, available online at: https://www.vdi.de/ueber-uns/ presse/publikationen/details/industry-40-technical-assets-basic-terminology-concepts-life-cycles-and-administration-models-english-version/ download?tx_vdipublications_publicationdetails%5Burl%5D=https%253A% 252F%252Fwww.vdi.de%252Fueber-uns%252Fpresse%252Fpublikationen %252Fdetails%252Findustry-40-technical-assets-basic-terminology-concepts-life-cycles-and-administration-models-english-version&cHash=818ab7752f2c 73b045fea7a35d3135e3, last accessed: 05.02.2024.

[19] Plattform Industry 4.0: DIN SPEC 91345:2016-04 Reference Architecture Model Industry 4.0 (RAMI4.0). Partner publication, in: Plattform Industry 4.0, available online at: http://www.plattform-i40.de/I40/ Redaktion/DE/Downloads/Publikation/din-spec-rami40.html, last accessed: 12.11.2016.

[20] Spinnarke, S.: OPC UA wird (neben anderen) Industry 4.0-Standard, in: Produktion. Technik und Wirtschaft für die deutsche Industry, available online at: https://www.produktion.de/trends-innovationen/opc-ua-wird-neben-anderen-industry-4-0-standard-334.html, last accessed: 05.08.2017.

[21] B&R Industry-Elektronik GmbH: Do you speak PackML? (B&R Industry-Elektronik customer magazine), in: *Automotion*, No. 05/2014.

[22] B&R Industry-Elektronik GmbH: Practical test passed: OPC UA becomes real-time capable (B&R Industry-Elektronik customer magazine), in: *Automotion*, No. 11/2016.

[23] B&R Industry-Elektronik GmbH: Safer in the line. System networking with openSAFETY (B&R Industry-Elektronik customer magazine), in: *Automotion*, special print edition 2012.

[24] Schröck, S., Zimmer, F., Fay A. and Jäger, T.: Konzept zur funktionsorientierten systematischen Wiederverwendung im Engineering automatisierter Anlagen der Prozessindustry, in: *Tagungsband EKA,* Magdeburg 2014.

[25] Schmertosch, T. and Kickinger, R.: Flexible drive technology for the production of individual products, in: *Tagungsband VVD* 18, Dresden 2018.

[26] Schöffel, J. and Gschrey, A.: Getränkeproduktion per Knopfdruck — Konzeptstudie "Bottling on Demand", in: *Anwendungen und Konzepte der Wirtschaftsinformatik,* No. 6/2017.

[27] B&R Industry-Elektronik GmbH: Automating lines safely (B&R Industry-Elektronik customer magazine), in: *Automotion*, No. 11/2016.

[28] Trützschler GmbH & Co KG: Textilmaschinenfabrik, available online at: https://www.truetzschler-spinning.de/, last accessed on March 8, 2018.

[29] Heinke, B.: Safety-related control systems. Implementation and application of EN ISO 13849 - 1, published by Phoenix Contact, Bochum 2009.

[30] Booch, G.: Objektorientierte Analyse und Design, Bonn 1994.

[31] Müller Martini AG: Perfect solution for edition 1, in: Panorama company brochure, Zofingen 2016.

[32] Kolbus GmbH & Co KG: Klebebinder KM 412 (product description), available online at: http://www.kolbus.de/produkte/anwendungen/klebe-binden/km-412/, last accessed: 05.08.2017.

[33] Liebau, D. and Heinze, I.: Industrylle Buchbinderei, Itzehoe 2001.

[34] Kipphan, H.: Handbuch der Printmedien, Berlin, Heidelberg 2000.

[35] eCl@ss e. V. Cologne: ISO/IEC-compliant reference data standard for the classification and unique description of products and services, available online at: http://www.eclasscontent.com, last accessed: 04.08.2017.

[36] B&R Industrial Automation GmbH: Online help of the B&R Automation Studio, Eggelsberg 2018.

[37]  Plattform Industry 4.0. May 2023. Progress report 2023: Industry 4.0: On the way to an intelligently networked industry. (Federal Ministry for Economic Affairs and Climate (BMWK), ed.) Retrieved August 20, 2023 from https://www.plattform-i40.de/IP/Redaktion/DE/Downloads/Publication/Manufacturing-X.pdf?blob=publicationFile&v=2.

[38]  VDMA e.V. April 11, 2023. Discussion paper — Interoperability with the asset administration shell, OPC UA and AutomationML: Target image and recommendations for action for industrial interoperability. Retrieved July 10, 2023 from https://www.vdma.org/viewer/-/v2article/render/78243357.

[39]  Industry 4.0 & IIoT, 14th edition July 27, 2023: IDTA publishes administration shell specification: Standardized digital twin in the industry. TeDo Verlag GmbH, Marburg 2023.

# Chapter 3

# Digital Project Planning of Machines

The previous chapter discussed in detail the role played by the specification of requirements for an automated machine concept and the knowledge work behind its development. This applies to the formulation of the customer's needs as well as the supplier's own business interests and the sustainability of the technical concept. The subject of this chapter is the realization of this specification as a ready-to-use technical automation concept. The development of this solution is by no means regarded as an open-ended research task. Prior feasibility studies, virtual functional models (the so-called mock-ups) and other techniques for property validation form the necessary basis for moving from the specification to a solution description. In Section 3.1, this idea is explored in greater depth and the importance of the modularization concept for the efficiency of this project planning is pointed out.

The V-model, which originates from computer science, is introduced as a central methodological instrument for project planning, which structures the computer-aided cooperation of different specialist disciplines in the development of mechatronic systems. Finally, this third section provides a concrete application description for the methodology of the V-model. This forms the basis for the coordinated use of hardware and software tools for the design of mechanical constructions, embedded electrical/electronic systems, their information-processing algorithms and, if necessary, other disciplines. Section 3.4 finally takes up our idea of the interdisciplinarity of modern design and production techniques in order to categorize their suitability for the future scope of functions in the dawning era of Automation 4.0.

## 3.1 Specification as the starting point for project planning

The decision to develop a new or further develop an existing production system is always of far-reaching importance and requires careful consideration even if the innovation cycles of the products to be manufactured often force this step. There can be two different starting points for a new development. A distinction can be made between the motivations of an "as-is" and a "to-be" approach. An *as-is approach* is based on an existing structure, whereby the target is formulated on the basis of a prior analysis of the initial situation and existing solution principles. Conversely, a target transfer from a defined ideal state can also justify a *to-be-oriented new development*, which entails a situation analysis subsequently.

The specification of a production system goes far beyond the formulation of objectives in the form of a requirements definition. The latter initially forms a basis for describing a common understanding of the performance requirements for the system to be supplied together with the customer or user. The solution-oriented implementation of the requirements specification agreed with the customer gradually leads to the definition or development of the solution concept by the manufacturer, whereby the manufacturer increasingly draws on its own proprietary know-how. In order to differentiate between these scopes of information and knowledge, a distinction is made between the provider's internal specification book of the supplier and the specifications that can be viewed by the customer.

On the way to a gradually increasing level of complexity and detail, *specification* gradually crosses the boundary into actual engineering, i.e. the actual design process. In this area, a methodology borrowed from software development has become established in the course of what is now a completely computer-aided approach. This is based on the procedure according to the V-model as a logical macrocycle, which will be discussed in detail in the following.

As part of the design process, it is sensible and desirable to use extensive work steps and work results on a largely recurring basis, irrespective of an actual as-is- or target state-oriented to-be-approach. As described in detail in Chapter 2, this is achieved by delimiting functional modules that can be reused without the need for new development. In addition to the flexible adaptability of existing solutions, the aspect of reusability of clearly definable functional modules is also an important motive for modularization. This book reports extensively on the extent to which the concept of modularization, which has long been established over decades, has received new impetus through the methodical penetration of the IT concept of object orientation.

A similar example is the standardization of the functionality of now Industry 4.0-capable components through the reference architecture model RAMI 4.0, whose layer-oriented structure is methodically based on the long-standing successful reference architecture of the ISO/OSI model for communication protocols. Its development began over 40 years ago and has since proven to be suitable for describing all generations of communication networks for technical systems. This shows how added values or requirements from the consistent application of the methodological principles, which are to be accepted as a necessity, develop a normative character. A communication module whose function is not strictly defined according to the ISO/OSI layer model can never be used reliably in such a standardized communication system. The demand for sufficiently reliable computer-aided predictability of every functional element in mechatronic systems so that it can be simulated as an experimentable model can be explained in a similar way. This is the only way to achieve property-assured project planning in line with an interdisciplinary procedure guideline.

Following this principle, a modularized I4.0-capable system can ultimately only be realized if all the modules that can be delimited within it have a digital function description that makes the overall system model (the digital twin) manageable and, in the best case, experimentable. Without a universally valid editing and management tool having been conclusively established yet, in the reference architecture model RAMI 4.0, the Asset Administration Shell (AAS) has been provided as the lowest level for this requirement (see Section 2.2.2.4). This is where the concepts of cross-domain modeling and co-simulation with a single tool currently meet on the one hand, which is the subject of this chapter, and the interoperable management and coupling (i.e. I4.0-compliant communication) of experimentable sub-systems on the other, for example, using the OPC UA communication standard [7].

## 3.2 Project planning according to the V-model

The V-model has proven its worth as a software development process for the increasingly detailed, cross-domain specification of a system with regard to its functionality in the course of project planning. The V-model is of great importance for the design process of automated production systems, which integrates various specialized domains. Even though this methodology has developed significantly for software production and has produced more powerful concepts in the meantime, the V-model mindset in the sense of a property-securing computer-aided design continues to have a firm place. This methodology was formalized in detail back in 2004 as part of the VDI guideline 2206 "Development methodology for mechatronic systems". The often inadequate equation with software design according to the V-model and the transferability to the so-called cyber-physical systems were reasons for a comprehensive revision of this guideline in 2021. This emphasized the logical sequence of the mechatronic design methodology [3, 4].

In order to explain the background to the relevance of this process model, some terminology must first be introduced.

### 3.2.1 Abstract and real model

To understand this, it is helpful to move away from the view that in computer-based design, some design steps are "initially" implemented with the help of computer support in order to prepare and support subsequent realization in a no longer digital "real world". It is much better to start from the perspective of a completely digitalized design and manufacturing process, of which the final physical realization inevitably brings the entire development phase to a close and gives it meaning. This approach already leads to the result of a *digital twin* [9]. This means that a computer-aided model is not only a virtual and possibly experimentable "model" of a real machine or system but will also remain its image throughout its entire life cycle. This approach is explained in Section 3.4 again and in greater depth.

From a methodological point of view, the physical-real (prototypical) execution of a digitized design is nothing more or less than its transfer from an *abstract* to a *real* model (Figure 3.1).

By definition, *abstract models* aim to provide an unambiguous description of specific properties of a system (existing or yet to be created) and are therefore also called *logical models*. This description is always based on a general and comprehensible syntax and semantics. In simple form, this can be, for example, a set of mathematical formula symbols, the graphic legend for a sketchy representation or a schematic diagram in the sense of a block diagram. This form of representation always

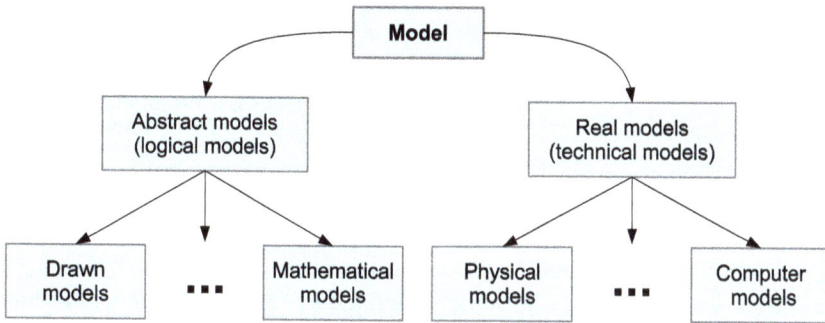

**Figure 3.1.** Distinction between abstract and real models.

serves the unambiguous (information) exchange between the experts involved, who arrive at the same idea of an artifact on the basis of the model. The RAMI 4.0 reference architecture model can also be understood as such a logical modeling by means of a complete, unambiguous functional description, whereby the following aspect must be included here (Section 3.2.2).

The focus of real or technical models, on the other hand, is their actual usability and feasibility. All tests carried out on such technical models are to be understood as experiments. If these experiments are carried out virtually, i.e. computer-based and numerically digitized, these experiments become simulations and have formed a third pillar of science alongside theory and experimental practice for decades.[1]

## 3.2.2 Model qualification, verification and validation

Thanks to the unambiguous nature of abstract models, they are now also suitable for use as implementation syntax for computer-based simulation. What is seen from the perspective of a simulation tool as a — possibly particularly intuitive — programming interface leads to a fundamental

---

[1] Both spheres described have existed since antiquity: We are familiar with historical mathematical treatises such as the mathematical formulation of the law of levers attributed to Archimedes as well as the traditional constructions of throwing machines and catapults based on them.

dissolution of the distinction between abstract and technical modeling outlined above.

On the basis of this differentiation, project planning as a commissioned design of technical systems is characterized by three phases:

(1)  The transition from the problem or task definition[2] to the solution-oriented design of the system is referred to as modeling or synthesis. The quality of the modeled system is expressed in whether it is suitable for describing a problem-adequate or task-appropriate solution design and is referred to as model suitability or model qualification. From the developer's point of view, an abstract model description is created for this purpose, which ideally can also be fully interpreted and experimented with using simulation tools.

(2)  To the extent that it is economical or justifiable when weighing up the additional effort and the remaining risk of failing to achieve the design objectives, prototype implementations of the solution concept are produced as an intermediate step. These should be suitable for experimentally testing the properties of the solution design. This step transforms the abstract model into a technical model. In the case of experimental model testing using virtual computer simulation, this transition from the abstract to the technical model corresponds to the programming of a simulator. In generalized terms, this is known as *(rapid) prototyping* taking place. The quality of a prototype is expressed as its *model verification* to the extent to which the relevant properties to be verified from the abstract model description are implemented without errors or deviation.

(3)  The requirements described in the task can be specifically checked for compliance in real or computer-based test setups using the prototype test model, as with later product samples. This represents experimental testing on a prototype instead of the final technical solution, which therefore corresponds to a simulation. The quality statement to be determined in this way is the *model validation or model rectification*. This refers to the evaluation of the real model properties to determine whether they are sufficient to fulfill the requirements of the task.

As a digitalized design chain, "shortcuts" are possible in the interdisciplinary design path of mechatronic systems if a distinction is made between

---

[2] The definition of the functionality that can actually be reproduced later on the real system corresponds to the specifications.

a path of increasing problem decomposition, initially conceived in descending order on the left, and a path of increasing solution integration, ascending again on the right from the highest level of detail. In order to illustrate this principle clearly, its symbolization with the letter "V" became obvious.

This simulation-based approach has all the fundamental advantages and disadvantages of experimental methods compared to formal or analytical property validation. In particular, only singular statements are produced in the context of the specific test scenario. Therefore, a strictly formalized and automated execution of this simulative testing is of great importance, so that a large number of tests carried out under precisely reproducible boundary conditions produce reliable statements that come close to analytical property validation.

Figure 3.2 shows these relationships between the real problem, abstract solution modeling and simulation of the design. The resulting triangle of relationships can be read as follows:

*If a prototype proves to be valid in the simulation experiments by sufficiently fulfilling the specified requirements, a qualified model, i.e. a suitable design, can be confirmed. It is always assumed that a verified prototype has been created on the computer through correct physical replication and, if necessary, programming.*

Conversely, if validation is not successful, the situation is not clear-cut: either the suitability (qualification) of the abstract model or the correctness (verification) of its prototypical implementation as an

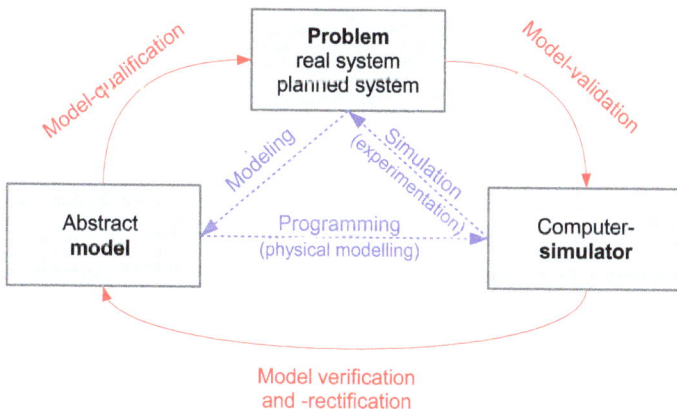

**Figure 3.2.** Conceptual relationship between qualification, verification, and validation.

experimentable model must be called into question. This explains why the term *validation* is used to describe both the individual step of simulative testing and the overall design evaluation cycle.

### 3.2.3 Computer-aided design

The special significance of experimental testing through simulation comes into play since not only humans but also computer programs can interpret the same description languages of abstract modeling. Originally, notation in higher programming languages was created for modeling logical processes explicitly for their evaluation and processing with computer programs (compilers). In the meantime, it has become ubiquitous that, thanks to the necessary unambiguity, any type of abstract modeling can ultimately also be interpreted correctly by computer programs.

By making the object of the description comprehensible and executable through computer evaluation, for example, by automatically processing a logical program notation or producing a graphically designed body shape by machine, the distinction between abstract and technical modeling described above is seemingly lost. A drawn block diagram, as a standardized abstracted representation for the purpose of documentation, is thus both the system-theoretical description of a control loop and its executable programming implemented in the graphical editor of a simulation system.

What is described as *computer-aided design* takes place precisely at the point where this interface merges: The synthesis of technical systems takes place in a language that serves both the design dialogue between experts and their correct documentation. As a model or system implementation, it is also the starting point for a computer-internal simulation or technical implementation with exactly the abstractly described properties. The outlined transition from the description model to the prototype, as well as the transition to the series product during further product maturation, may be fully automated, whereby the correctness or "verifiedness" of this prototype is generally not called into question during automatic transfer. However, compliance with all predefined conditions of use in accordance with the computer-aided design as well as for its prototypically generated test sample[3] remains the indispensable responsibility of the developers within the framework of an expert approach.

---

[3]For example, the stable numerical calculation of differential equations must be guaranteed.

Mechanical, electrical/electronic and IT design of products or production systems, including other disciplines involved, thus produce their results as a digital image as a result of the computer-based development outlined above, which in each case contains a solution description in a suitable modeling syntax that meets the requirements. As described, these designs are characterized by duality: In their form of a description language, the designs function as an abstract model and in their notation, which can be automatically processed by the computer-aided design tool into an experimentable system, they form a technical model.

Finally, with regard to simulation, it must be noted that a complete match between the real system and its computer-based model is fundamentally unattainable. On the one hand, every model emerges from the description of a real system via abstraction and idealization. It therefore contains far fewer depicted effects or influencing variables and for this reason will be idealized and will behave differently in principle. On the other hand, any measurement of a physical variable on or in a real system, be it a prototype or series product, is associated with measurement errors, which generate deviations between the simulated model result and experimentally determined comparative measured variables. A match between the real system and its computer-based model or prototype is therefore only possible within a specified tolerance. Thus, model validation is particularly difficult if or as long as no measurements are possible on the real system. This is the case for systems that are difficult to access or are still in the planning and development stage, as well as for rare or undesirable system states. Closing these gaps is also the special significance of the logical approach according to the V-model in order to experimentally test prototypical realizations of preliminary or incomplete design statuses at an early stage.

In the information processing sector in particular, however, the tools and programming languages directly intended for the purpose of implementation can themselves be used for design development instead of the abstract modeling step. This will later also include the orchestration of an executable and therefore experimentable overall system consisting of I4.0 components. As far as possible, however, transferable high-level languages should be used above the system implementation. Otherwise, the reusability of the implementation code is limited, which not only causes considerable transfer effort but also harbors a significant source of transfer errors that are difficult to narrow down.

### 3.2.4 Modeling variants

From the perspective of the systematics of Figure 3.2, the approach of designing for the functional level of components directly in the implementation language seems as strange as designing a mechanical component without any geometric drawing solely on the basis of the associated machining program of the machine tool used.

The European standard EN 61131-3:2014-06 is a globally applicable standard for programming languages for programmable logic controllers (PLCs) that are used to implement sequential control sequences in manufacturing plants [1]. In this standard, five languages are defined: the *instruction list, contact plan, function block diagram, Sequential Function Chart* and *Structured Text*. Each one enables the implementation of logical behavior according to its own tradition. Within this standard, only the so-called Sequential Function Chart (SFC) is suitable for the abstract modeling of logical sequences. It lays the foundation for graphical modeling using state diagrams as finite automata, which have already been introduced in Section 2.1.3. In these graphs, the places represent different states, which are connected by (conditional) transitions in the form of directed edges. On the basis of this graphically representable form of description, a consistently abstracted modeling is carried out independent of the later realization as a technical model. At the same time, this modeling enables seamless integration into engineering and simulation tools based on graphical editors. However, a consistent way of thinking in this process language in no way forces you to dispense with text-based programming using a high-level language. Object-oriented programming is sufficiently powerful to allow systems of states and their transition conditions to be expressed using high-level languages by means of appropriate libraries. The libraries for modeling step chains or sequence controls using state diagrams therefore represent an extension by an additional processing form for the design tools to configure the information-processing functions.

Initially, and still widely used today, step chains or sequence controls were designed solely for discrete-time information processing by digital computers. In programmable logic controllers, this was done by processing a cyclical program in a continuous endless loop. In order to model the processing of continuous signals and dynamics also, the signals must be sampled discretely in time and preconfigured accordingly so that they can be processed by cyclic algorithms. The continuous transmission behavior is modeled by ordinary differential equations, which are solved numerically according to the discrete-time step size, with each calculation cycle corresponding to a virtual time slice.

Modeling an information processing system as an experimentable prototype requires real-time simulation. This means that the virtual time slices must be processed by the executing computer at exactly the same time interval as they were calculated by the numerical solution method. As the complexity of the model and the temporal resolution of the simulation increase, the computing power of the simulation system must be increased accordingly so that the calculation results are available throughout the simulation before the end of the respective time slice.

Finally, with the state charts representing process language, there is also the form of *discrete-event simulation.* Their execution in the computer is generally not actually event-controlled by the interrupt system, which in practice is largely reserved for handshaking with real-time-critical peripheral modules. Instead, Chapter 4 describes the criteria for cyclic sampling of input signals, which are also used to control the transition conditions of the event-dependent state models. In this way, all discrete-event models can also be modeled and simulated in discrete time. Key requirements for real-time capability are discussed in more detail in Section 4.2.

A clear distinction between the process of developing logical sequence programs in the course of the engineering process and genuine prototype testing appears to be difficult but is exemplified by the use of emulators. These are systems for the targeted imitation of the information processing of specific target hardware. A distinction is made in particular between software emulators for emulating program execution on a specific CPU and, conversely, the physical emulation of the CPU in its socket by means of a so-called *in-circuit emulator.* These two variants again reflect the duality between the left branch of increasing detailing and the right branch of increasing integration of a design based on the V-model. The following section is dedicated to its introduction.

## 3.3  V-model in the application

The previous section should make it clear that computer-aided design is generally the key to an approach based on the V-model. The idea of this model is that in a logical process model, from reaching the design implementation, each realization step corresponds to a certain intermediate stage of integration, i.e. the increasingly completed realization from sub-elements, along an ascending right-hand branch of a simulated "*V*".

### 3.3.1 Basic structure and property protection

The correspondence of the two branches shows how, on a common level with the substructure created on the right-hand side (component/module verification, system verification), precisely those properties can be checked that were brought about in the specification-compliant design step on the left-hand descending branch of the decomposition (component/module synthesis, system synthesis) (Figure 3.3).

The horizontal arrows in Figure 3.3 are used to characterize the property validation within a common design level. It should be noted here that the verification of modeled substructures of the overall system is as detailed as it is mapped at this level of the requirements specification. As long as it is not a matter of testing in a real application environment at the highest level, we can only speak of a *verification* of the properties created in the design. Only at the level of a fully integrated system, practical testing as a sub-component or complete system is possible, which corresponds to its *validation* in real use.

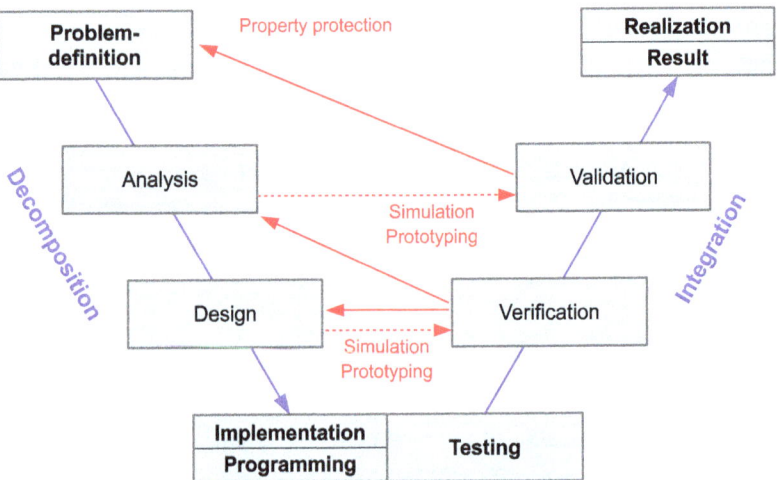

**Figure 3.3.**   Abstracted V-model: functional overview.

It is very important to reach the highest testing level of overall integration as early as possible in the design process with the help of prototypes.[4] It cannot be ruled out that after successful verification of all sub-systems, the overall system will still fail in its practical testing, i.e. that validation will not succeed. In this case, the cause is as obvious as it is sobering: already the definition of the requirements specification was faulty or incomplete. It is in the nature of such errors hidden in the specification that they occur completely unexpectedly — if they had been expected, they would have been taken into account in the requirements definition long ago.

The increasing importance of a specification with as few errors as possible is not only the result of computer-aided design, which reliably translates this specification into a corresponding technical solution. Rather, the demands and requirements for the development of new systems are increasing to be no longer evolutionary depending on predecessor types in their *as-is state*. A variety of concerns, such as the use of new production technologies, the integration of flexible modularity, resource-saving adaptability to changing production outputs or simultaneity with other development stages that have not yet been completed, are increasingly leading to completely *to-be state-oriented new developments* that can only be partially tested and further qualified on the basis of already established products.

The first flight of the Airbus A380 in the year 2005 is an anecdotal example of this. One could almost believe the media excitement surrounding this event as if it had been about whether this aircraft would actually fly. However, virtual prototypes had already proven themselves millions of times on the way to their maiden flight — from flight simulators to structural load tests. What really matters during the first flight of new aircraft models within their development process is establishing the precision of the simulation models used to date and the corresponding adequacy of the requirements specification. Yet, it could happen that in the specification or modeling of such a large aircraft, for example, the design of the particularly long cable harnesses in the compartments between the aircraft frames could be underestimated. Since it was only realized during final assembly (also due to coordination problems between the production sites) that plug connections could not be reliably closed because the cables

---

[4]For example, visualizing mock-ups or 3D-printed construction samples.

were too short, this contributed to a delay in deliveries of over a year [2]. No simulation can reliably detect such errors as long as they are not expected and their causes are therefore not mapped accurately enough in the specification and modeling.

## 3.3.2 Decomposition

With increasing detailing at lower design levels, *decomposition* can be understood in two ways for production engineering and mechatronic systems: On the one hand, it refers to the decomposition into sub-modules and sub-sub-modules, as already comprehensively described by the concept of modularization. On the other hand, this subdivision also enables the problem to be broken down into different specialist domains involved in the design, which provide the mechanical, electrical/electronic and information processing functions. The *V-model* provides the methodological basis for these specialized domains to be run through in parallel in an integrated design process. The original VDI/VDE guideline 2206:2004 "Development methodology for mechatronic systems" already focused on this logical macrocycle as a central integrating concept (Figure 3.4). In the updated version of the guideline "Development of mechatronic and

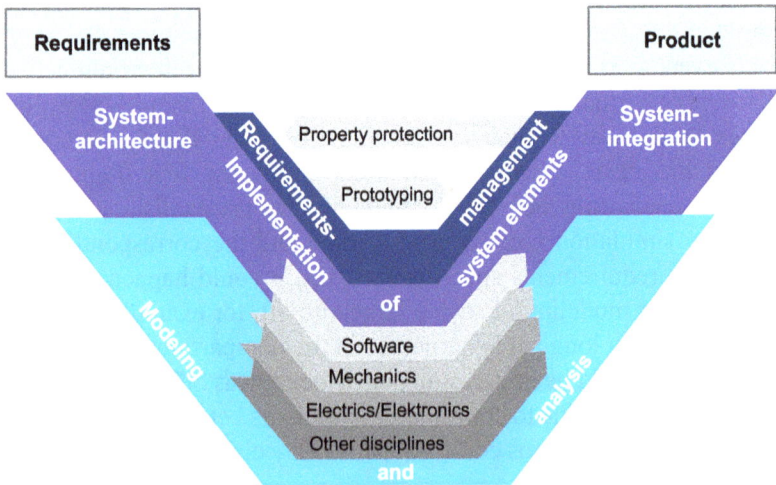

**Figure 3.4.**    V-model as a macrocycle of mechatronic design (According to [3]).

cyber-physical systems", it was emphasized that *mechatronic systems* do not necessarily have to be an exclusive triad of the three specialist domains mentioned above, but that other disciplines can be easily integrated.

The differentiation of rapid control prototyping (RCP, left branch) and hardware-in-the-loop simulation (HIL, right branch) on the basis of numerical simulation of a dynamic system (Figure 3.5) originates, for example, from the information and signal processing domain. However, the same is easily transferable to other domains. From a mechanical point of view, constructive designs (left branch) can be visualized and experienced through visual animation (mock-up/augmented reality) in the form of virtual reality. This allows ergonomic properties to be checked as well as fundamental questions of maneuverability and mountability. Mechanical realization of (preliminary) sub-systems for testing prototype samples (right-hand branch) is made possible by rapid prototyping technologies, for example, by providing preliminary one-off products at a justifiably low cost and time. For decades already, the world of electronic systems has been familiar with editors for electronic circuits and circuit boards with integrated function simulation (left branch) and the possibilities for their fully automated sample production in small series (right branch). Here, the products differ significantly less in their properties between

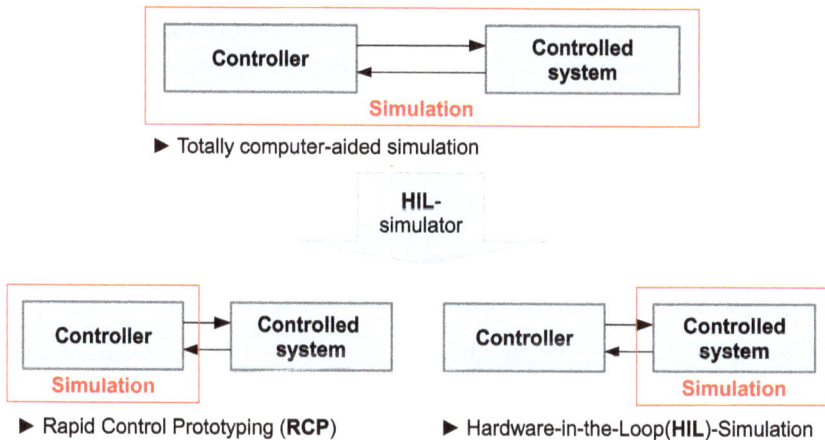

**Figure 3.5.** Comparison of rapid control prototyping with hardware-in-the-loop simulation.

prototype and series production than with 3D printing of mechanical prototypes.

### 3.3.3 Modularization and object orientation

A commonality becomes apparent, which combines the nature of rapid prototyping of mechanical components, for example, using 3D printing, with the automatic production of sample boards and the real-time simulation of information processing systems. This similarity consists in the fact that the right integration branch of the V-model is achieved directly at a selected level of detail of the design with a manageable material and, in particular, time expenditure. Without having to go through the entire design process, including implementation and production design, a functional model can be completed into a functional prototype. This step in particular enables early application testing and thus helps to recognize errors in the specification. In view of the systematic quality assurance during computer-aided design through comprehensive verification, any error in the specification would otherwise inevitably only become apparent as a corresponding malfunction in the near-series product.

The consistent specification-compliant assurance of properties also includes another fundamental level of significance. Prototypical, partly simulative validation, even without complete implementation of certain system areas, also integrates the concept of modularization in this way. Different degrees of implementation or product maturity can be combined in a side-by-side arrangement of different modules and components. In an experimentable overall simulation, it can be determined for each module whether a development sample, a series component or a pure simulation is used. The basic structures occurring along the development progress are differentiated in the following section.

### 3.3.4 Basic structures of simulative testing

From the point of view of automation technology, the step toward a process model that can be applied across domains is particularly easy to understand using the example of the simulation of dynamic systems. A widely used software package for this purpose, *MATLAB/Simulink*[5]

---

[5] Manufacturer: The MathWorks, www.mathworks.com.

with its graphical editor for model implementation, is closely based on the notation of signal flow diagrams using formalized block diagrams. The orientation toward an established technical concept is certainly one of the reasons why this software is widely used, particularly at universities, and has been the source of numerous application impulses [5]. In the industrial context, other simulation tools such as *LabView*[6] are also well established and are now suitable for the procedure outlined here from a design to an experimentable system [6].

The standards based on block diagrams form the framework so that their directed graphs form an abstract model that can be interpreted without errors. They consist of signal arrows as their edges and (rectangular) blocks as their transmission elements (nodes). Ultimately, the graphical editor of the design tools is already used to implement a dynamic system represented by the block diagram as a technical model in order to carry out virtual experiments in numerical simulations.

Since its introduction, a special additional library within *MATLAB/ Simulink* has become increasingly important. Its original name, *Realtime Workshop*, reflects the fact that the full scope of its significance was by no means foreseeable at first. The initial intention of this Simulink extension was to utilize external target systems for the execution of simulations as real-time applications. At the beginning of the 1990s, the aim of such external high-performance computer systems was to use model simulation as a stable real-time application, possibly coupled with real input/output interfaces. In particular, accelerated hardware platforms (transputer/DSP/ RISC processors) were used for this purpose, which offered significant speed advantages over personal computers and workstations at the time. This tool is now aptly referred to as a *coder.* In general, this library can be used to export simulation models from Simulink as executable source code. This source code in high-level language can then be used for any target system for which the corresponding compilers (preferably C/C++) are available. Today, this interface is as important for the design chain of information-processing functions as the CAD/CAM interface is for the development chain of geometric design programs: the execution code can be generated with the design tool fully automatically for the production of prototypes through to industrial series production.

The production of suitable test systems for the real-time execution of software applications, which emerge directly from the implementation in

---

[6] Manufacturer: National Instruments, www.ni.com.

the simulation system, has long ago developed into a customized industrial branch of business. These systems are known as *hardware-in-the-loop (HIL) simulators*. This means that, thanks to high computing power and real signal interfaces, it is possible to couple in real time the transfer behavior of real assemblies, i.e. modules or components, for parts of this simulated circuit (loop) within the complete signal loop consisting of the modeled controlled system (i.e. the model of the system to be controlled) and the control algorithm (controller) to be designed for it. Within this device technology, known as *HIL simulators*, a clear distinction must be made by the type of application. Only those variants in which a real control component is coupled with the simulation of the controlled system correspond to the so-called HIL simulation in the narrower sense. After all, only with these is it possible to test the control system experimentally in all specified situations from the system simulation. In all those cases in which the designed control algorithm is first connected and tested experimentally with the real modules of the controlled system by real-time simulation, this is referred to as *rapid control prototyping (RCP)* (Figure 3.5).

The distinction described here is based on the fundamental differentiation in terms of the V-model. RCP marks an intermediate step in the increasingly detailed specification of the system to be designed; HIL, on the other hand, represents an intermediate step in property protection on the basis of prototypically functional modules. Figure 3.6 outlines this distinction by assigning RCP and HIL to the two opposite branches of the V-model.

As a project management method for software development, the V-model was originally established with a much higher level of detail, as shown in Figure 3.7 in simplified intermediate steps. Design steps from the information processing domain are assigned outside the boxes. At the same time, it should be pointed out that the processes discussed here as examples are by no means as clear and delineated in the practice of mechatronic designs as the structure of the figure suggests. Rather, the diagram forms the orientation framework so that the design process can be structured logically. Each design step can thus be categorized in this process model independent of the development tool used. Various examples of this are presented at the end of this section.

For a long time, the methodology described above was used consistently, especially for complex mass-produced products such as automotive assemblies. Accordingly, primarily HIL systems with all conceivable interfaces used in this automotive sector are available on the market.

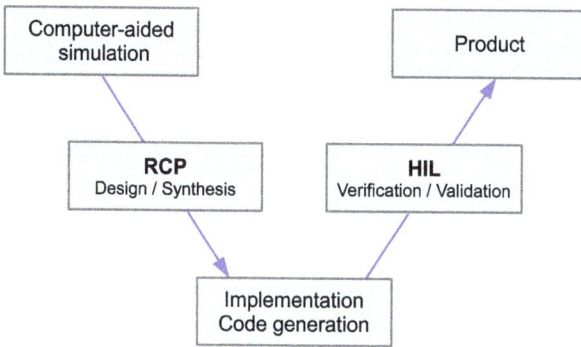

**Figure 3.6.**   Development steps where RCP and HIL are assigned to the branches of the V-model.

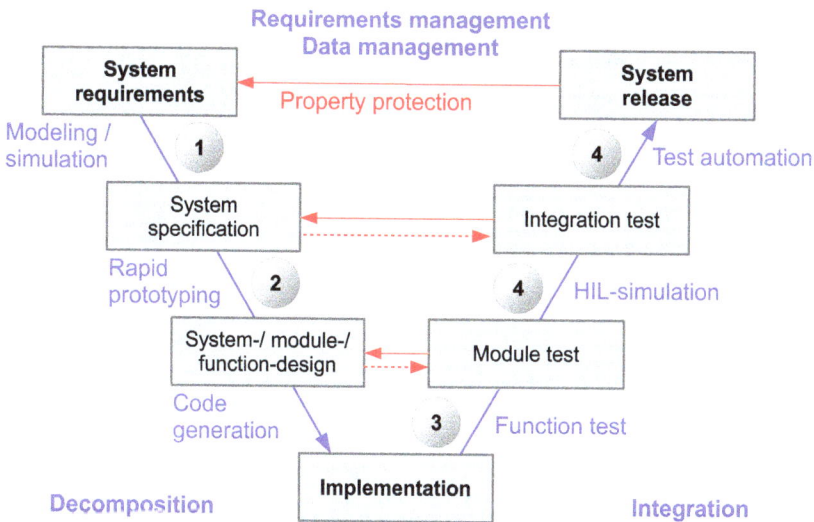

**Figure 3.7.**   Detailed representation of the V-model and assignment of the validation methods.

The situation is quite different in the area of production systems. As a rule, this involves very small series of identical products that have not justified the expense of complete computer-aided modeling for a long time. In addition, the *as-is-oriented approach* is generally used here, i.e. a more evolutionary further development of previous versions, which are at least as suitable as a prototype for experimental development work as a virtual model. As a consequence of a virtually non-existent need, the communication interfaces commonly used in such production systems in HIL systems were not available at all for a long time.

This situation has now changed: Modern control systems not only have computing power on par with HIL systems, but they have also programming interfaces for automated code generation from simulation systems. This means that these industrial controllers are suitable for rapid control prototyping or can be used as a real-time simulation system for a dynamic plant model to be a HIL simulation system. At the same time, the input/output communication of today's control systems is mapped by virtual interfaces anyway, which can easily be redirected as a data stream to simulation systems.

These new possibilities are also meeting with correspondingly growing demand: to ensure that the development of production systems can continue to keep pace with the requirements for flexibility, technological progress and decreasing lead times, completely computer-based, mechatronic design concepts are now also indispensable here. Practical advantages for commissioning have also made it a standard practice very quickly to test control system designs extensively in advance by simulations in a simulation environment created for this purpose that is similar in function to the later system.

The question now arises as to which example can be used to illustrate such a run of the V-model, as shown in Figure 3.7 can be reproduced in a representative manner. It must be understood that actual practice tends to follow such methodological guidelines piecemeal. Rather, the V-model serves as an orientation framework for assigning the development steps that actually take place to this prototypical process pattern.

With regard to the example of the mechatronic model *glue station* of an *adhesive binder* introduced in Chapter 2, the first step in this context is to model and simulate the technological process, i.e. the heating and application behavior of the adhesive used. A behavioral model of this process based on system analysis, for example, in *MATLAB/Simulink*, forms the starting point for the design and implementation of all control functions. The control design can then be completely executed in a simulation system, such as *MATLAB/Simulink*, coupled with the controlled system simulation. This phase is referred to as standard *model-in-the-loop simulation*, which can be used to drive forward the specification-compliant implementation of control functions on a model basis (Figure 3.8).

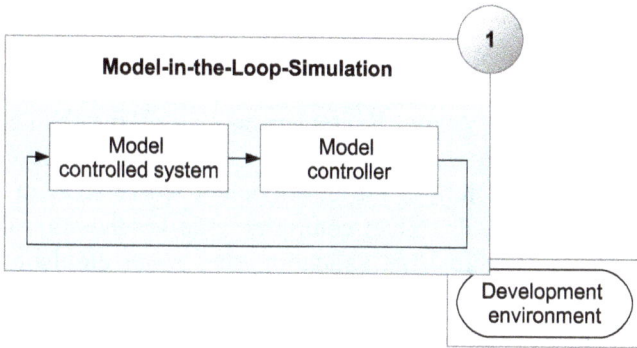

**Figure 3.8.** Model-in-the-loop simulation: The route model and controller design are fully represented in the simulation tool on the development PC.

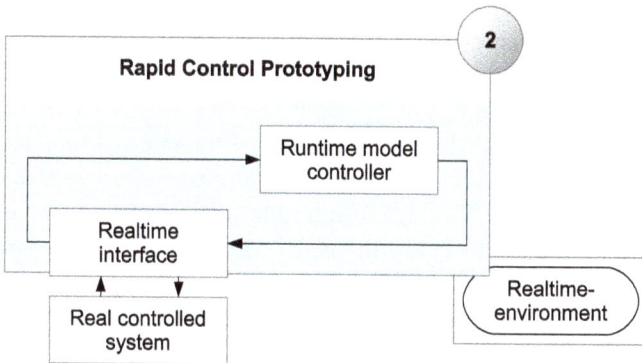

**Figure 3.9.** Rapid control prototyping by coupling the real controlled system with a real-time system that executes the control design.

As soon as a design with specification-compliant simulation behavior is available, it is possible to move on to real-life testing of the control design. This requires a real-time controller simulation system that can be coupled with the sensors and actuators of the real system (Figure 3.9). As already described, thanks to open programming interfaces and higher computing power, some OEM controllers can now also be used as suitable simulation hardware for the runtime code automatically generated from *MATLAB/Simulink*. In the example of the glue station, the adhesive heating is one of the actuators to be controlled, as is the handling of the adhesive application. In the real-time system, the real sensor signals of the process, which were simulated during a model-in-the-loop simulation using a behavioral model, must then be able to be measured in order to monitor the correct execution of the control design.

In a further design phase, the implementation on the final controller hardware must be prepared based on the tested control prototype. To do this, we return completely to the development workstation in order to automatically generate the target controller code for the controlled system. The prerequisite for this validation step is an emulation of the intended control architecture, which can be coupled with the controlled system simulation (Figure 3.10). This step is known as *software-in-the-loop simulation.*

The automatic generation of the control code from the design model is of central importance in the V-model mindset. This is not only for reasons of efficiency. Supposed errors in the control design, which have not yet been recognized as such during validation, represent a smaller source of errors later on than errors that creep in by chance during any kind of post-implementation of the control model for the final target architecture.

For a completely abstracted modeling of the control functionality in the preferred form of a graphical sequence language, the *StateFlow* blockset is available for the MATLAB/Simulink design system, for example. The transition to a controller with an EN 61131-compliant programming interface then requires a way of seamlessly transferring a state transition diagram from *StateFlow* to the corresponding *sequential function chart.*

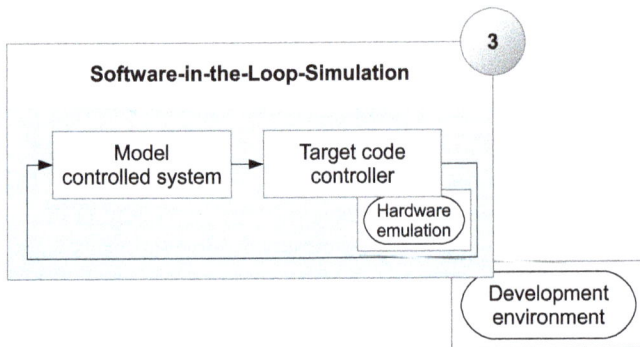

**Figure 3.10.** Software-in-the-loop simulation of the control implementation coupled with the route model.

This transfer is automated at least at the transformed level of the corresponding high-level language source code in the syntax of *Structured Text*. This is also methodologically acceptable, provided that every change and correction to the control algorithm continues to be made in the StateFlow editor of the state diagrams.

Last but not least, the HIL simulation is used to comprehensively validate the control system by its preliminary coupling — as it is intended for installation in the production module — with a real-time simulation of the system (Figure 3.11). This method allows the risk-free simulation of all conceivable system states and error situations at a reasonable cost, so that the adequate reaction of the control system can be checked. In the example of the adhesive binder, the previously used controlled system model is coupled with the intended module controller as a real-time simulation. Commercial HIL simulators can be used as hardware as well as PCs that have been configured as a corresponding *real-time target machine* and are equipped with the necessary real signal interfaces. Another option is powerful industrial controllers that can also be operated as a *real-time target*. Based on such test bed, there are no limits to the simulation of error situations, such as sensor failures, heating failures or an overflow or exhaustion of the adhesive supply.

The latter application in particular, HIL simulation, has become considerably more important in recent years. The reason for this is the possibility of thoroughly validating the intended control system at the developer's workplace without the aggravating external conditions of the

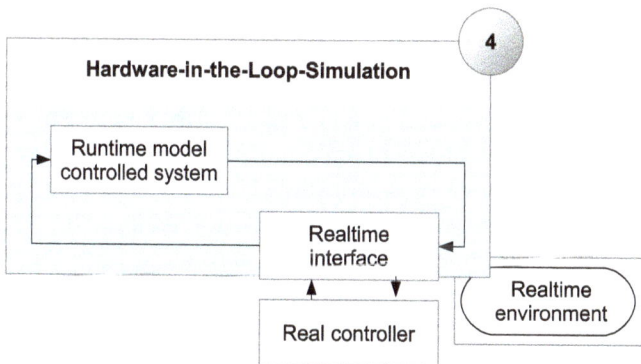

**Figure 3.11.** Hardware-in-the-loop simulation by coupling the control module with a real-time simulation of the controlled system behavior.

subsequent installation site (time pressure and remote construction site, incomplete system operation, non-reproducible errors). This trend is supported by advanced simulation tools that can be used to model and simulate production processes such as intralogistics sequences and workpiece modifications with reasonable effort and coupled with a control system, which is an integral part of an "Automation 4.0" approach in terms of the subject matter of this book anyway. In addition, thanks to open programming interfaces and high computing power, control units can now also be used quite easily as a real-time simulation system.

In Figure 3.7, all four forms of simulation-based testing in the development process are located in the V-model. In each of these development steps, recourse to such a simulation-based methodology is a selectable option. This is always associated with additional expenditure on additional technology and for modeling and implementation. More and more frequently, however, recourse to these methods is justified by the preconditions of the innovation cycle. At the same time, this explains why it is necessary to decide which of these steps is appropriate and should be included in each new development process.

## 3.4  Transferability of the interdisciplinary mechatronics approach

The previous sections have outlined how the process concept of a mechatronic design is suitable both for dealing with the inherent technical interdisciplinarity and for the sustainable implementation of a modularization strategy. The mere existence of mechanics, electrical engineering and information processing does not constitute a mechatronic system. Similarly, the reuse of assemblies alone does not result in a modularized concept. To put it in a nutshell, it is not necessarily possible to determine whether an automated component or complete system in its (series) manufactured state is truly *mechatronic* or modularized or not. This categorization is essentially determined by the previous process for its design, testing and integration. In the case of a mechatronic system, the complexity of the design has made it necessary to work on the task in an interdisciplinary network of mechanical, electrical and IT domains. At the same time, a targeted decomposition into delimitable and interchangeable functional levels leads to modularization with planned reusability and more manageable complexity. The fundamental added value of this approach lies precisely in this interlocking, as otherwise the mechanical design, electrotechnical equipment and IT commissioning of the overall system have to take place in separate, consecutive steps. Put simply, the mechatronic process model helps to master the complicatedness of technical systems, while modularization helps to reduce their complexity. Highly integrated mechatronic systems such as dual-clutch gearboxes in cars are impressive examples of how their degree of functional integration is inconceivable without a cross-domain, decomposing and integrated design process.

### 3.4.1 Simulative testing of large systems

By simultaneously turning the design tools into implementation tools for prototypes through to series maturity, abstract modeling and validation through simulation merge into an integrated step at component or module level. However, an extended scenario is required for the transition from individual components or modules to an overall system, which is represented in the sense of Industry 4.0 by an orchestration of *Experimentable Digital Twins* (EDT). Instead of a bloated, integrated overall simulation, a coupling of systems is pursued on the basis of their external representation in the form of the so-called Asset Administration Shell (AAS). This component model is the result of a further implementation step, which also becomes the subject of its own functional verification. As a result, an experimentable overall system can be composed of experimentable components or modules whose interconnection is based on the Industry 4.0-compliant documentation of functionality and communication by means of their administration shell [10]. A distributed co-simulation of several modules as an overall system would inevitably include modeled and simulated communication behavior between the sub-components. Instead, the coupling of various real or simulated modules in a HIL-like approach can already be used for the realtime communication system finally intended for overall integration (Figure 3.12) [11].

The advantage of this approach is that, just as the system implementation code is already part of the simulation and therefore evaluation at component level, this now also applies to the communicative coupling of the administration shells of I4.0-capable modules at system level. This means that the modeling of communication and functional interaction becomes part of the experimentable overall system and validation.

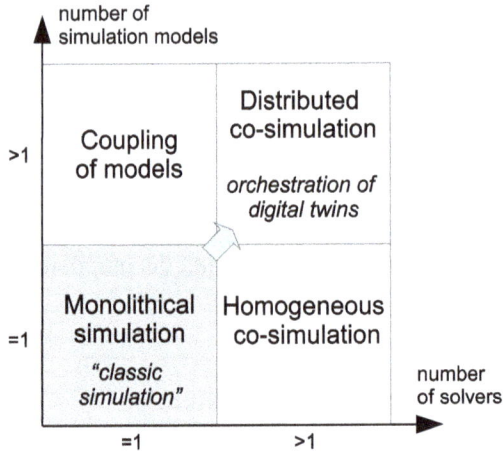

**Figure 3.12.**    Extension of the simulation-based design scenario for orchestrated systems of digital twins (simplified according to [8]).

Figure 3.13 shows how harmoniously the scenario of the co-simulation of experimentable digital twins according to [8] fits into the concept of the V-model.

Two challenges for the approach described must also be mentioned at this point, which is why it is currently more of a research field than a "state-of-the-art" concept. First, the necessary encapsulation of each I4.0-capable component in its own asset administration shell is currently an additional expense that will only be gradually reduced — be it in any one of the following scenarios:

- as more and more components already have these capabilities;
- by certain customers explicitly demanding this capability despite additional costs, or;
- by using tools and routines to make this effort manageable over time.

Second, simulation model coupling via I4.0-capable communication is not an exhaustive solution for a simulatable overall system. Physical interactions take place between the coupled sub-systems, which must necessarily be modeled as an experimentable twin, but do not contribute to the creation of the functional coupling in the projected communication channel itself. This communication system is anyway not designed to fulfill the possible requirements for the exchange of signals for the

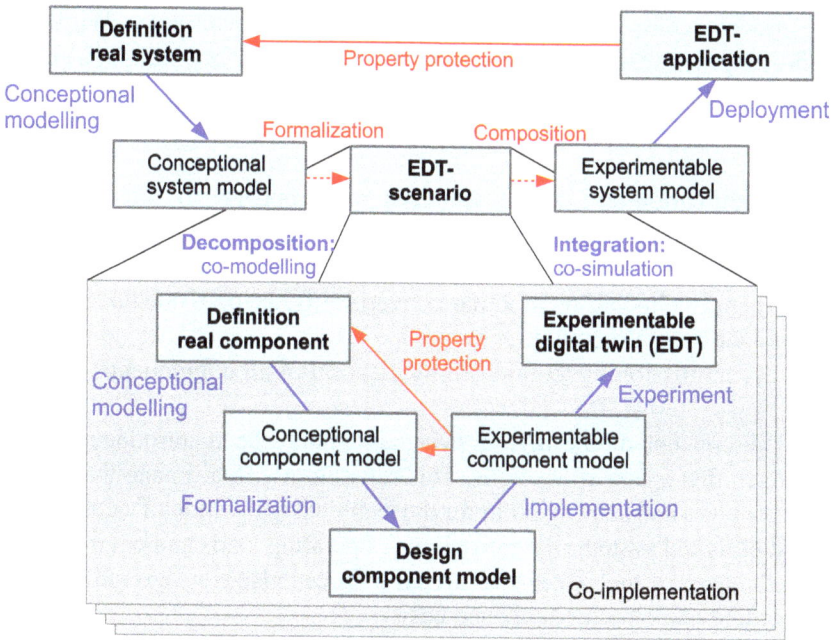

**Figure 3.13.** Experimentable digital twin through decomposition at system level for a co-simulation at component level.

coupled simulation of these module interactions. For this reason, a separate methodology for physical model coupling in a comprehensive "metaverse" will first have to be established.

## 3.4.2 Life cycle modeling

So far, we have assumed that the computer-aided design process of a technical system always assumes its ideal state. This means that the system is idealized during the design process by assuming that it is as good as new when it is commissioned, as this is exactly what the computer-based design model represents. However, every technical system goes through a life cycle over its useful life. Production-related faults lead to deviations from the time of commissioning, as does aging due to use and environmental influences. Countless new possibilities are currently emerging for the unlimited generation of measurement data over the entire life cycle of the production plant from seamless quality monitoring of production by the process control system to distributed sensor technology. With this approach, the design concept is based on the asset of a production plant[7] and will need to be expanded in future to include the dimension of the plant life cycle, which comprises the following phases:

(1)  specification and design, finalized with approval by the client and authorities;
(2)  construction, completed with commissioning and testing;
(3)  operation, including modular convertibility, accompanied by maintenance and retrofitting;
(4)  decommissioning, completed by disposal of all components.

The concept of the *digital twin* has already been introduced as an approach that spans all sections. This refers to a virtual image that is not only the basis and starting point for the technical realization of a computer-based designed system. By recording all operating loads and environmental influences of the system, its digital twin can also run through the rest of its life cycle on a model-based basis.

However, in the context of existing automation practice, the entirety of the steps has so far only been the subject of an integrated consideration in the context of computer-based design to a limited extent. By linking the second and third phases according to the asset view, (at least) three specialist domains that have previously dealt with technical systems independently of each other come together similar to the interdisciplinary origins of mechatronics:

(1)  the operational management of an active production plant within complex value chains, depending on market requirements, profit targets, resource availability, etc.;
(2)  the technological feasibility of the utilization program to the production plant with changing production targets, including constructive changeability through modularization methods;
(3)  condition-oriented and preventive maintenance to ensure system availability.

The conceptual integration of these domains will be a key aspect of the future challenges for automation design in the Industry 4.0 era. From

---

[7]See Section 2.2.4.

today's perspective, this step appears to be just as plausible in its approach and objective, on the one hand, and still unstructured in its methodological implementation, on the other, just as it was the case three decades ago for the emerging integrated approach to mechatronic systems.

Looking back, an integrated view of components with mechanical, electronic and IT functions as a complete mechatronic system provides us with a comparable figure. In the 1990s, mastering the associated complexity appeared to be a plausible vision, for which an engineering methodical approach was still conceptually undeveloped. It was only gradually that a concept was developed for the domain of *mechatronic systems* that did not primarily establish a new cross-sectional discipline, but rather a suitable cross-domain methodology. This approach, described in the previous section, was formulated as a guideline [3] in German-speaking countries in 2004.

### 3.4.3 Limits of simulation-based evaluation

As an extension to the concept of the V-model, a key approach for the design of assets will be to ensure that components reliably fulfill pre-specified functions in their interconnection for the life cycle even if all of these complex situations cannot be fully run through experimentally in advance using simulations. It will therefore be necessary to trust in the functional correctness of the agreed specifications even without exhaustive simulation-based testing, just as the V-model trusts in the correctness of automatically generated prototypes. This will be all the more the case if mechatronic systems in the sense of intelligent components or adaptable modules utilize integrated information processing not only for the production of integral functions but also for flexible adaptation to the requirements of their embedding or use. The systems are thus programmable in their function and become an independent interaction partner of their environment not only physically but also in terms of information technology. By anticipating and providing information processing capabilities to such an extent that these systems can be actively represented by a digital image, they develop not only as an overall system but also in their sub-modules into the so-called "cyber-physical systems" (Figure 3.14).

In order to cope with this development, the process model of design will approach the general methods of software design, which makes a V-model, which is strongly simulative or experimental, appearing to be a

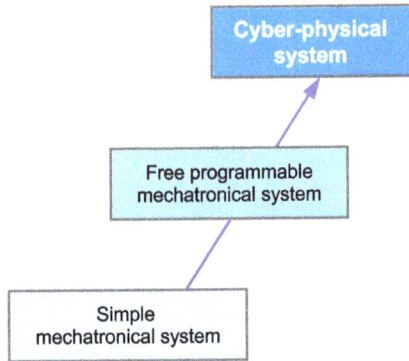

**Figure 3.14.**  Historical development from a simple mechatronic system to a cyber-physical system.

special case. In particular, the functional interactions with the physical world require experimental testing in the V-model. For the increasingly frequent interactions between digital systems, on the other hand, the correctness of the implementation in accordance with the specification is sufficient and therefore decisive. Formal methods are best suited for checking the completeness and absence of faults and contradictions of the interactions to be specified, for which numerous example publications were mentioned in Section 2.2. Nevertheless, test simulations remain an indispensable part of validation here too. In addition, these specifications always include a defined handling of error states. The avoidance of undefined states is easier to ensure for reliable internal interactions of digital systems than a guarantee of total freedom from errors.

It should therefore be pointed out at this point that the limits of V-model-based engineering become apparent: Only a certain class of models of physical systems with limited complexity is suitable for real-time simulation. Robotics can be used as an example to illustrate this: On the one hand, the time scopes of handling processes in industrial production are typically suitable for analysis in real-time simulation. The time periods to be analyzed, which are covered by the production step for a single product, can be easily mapped using a closed simulation cycle. On the other hand, simplified dynamic models are generally used here. The kinematics of an industrial robot are abstracted as a simple multi-body system with point masses and rigid motion elements. Due to the available computing power, it is now also conceivable that a real-time simulation is carried out as a so-called continuous system. This means that elasticities within the structure are also taken into account. Comparable to the assumed point masses in the centers of mass, this flexibility is also considered in a simplified manner using virtual spring elements in the joints.

In fact, the elastic deformations of the structural elements (as well as thermal or age-related deformations) occur spatially distributed over the entire rigid body. The *finite element method* (FEM) is also an adequate modeling method for this, but its implementation in the context of a real-time simulated prototype is illusory for the foreseeable future in terms of the required implementation and, in particular, calculation effort.

Ultimately, the outlined context dependency of simulation results is symptomatic of a disadvantageous aspect of the described simulative design evaluation according to the V-model procedure. It must be pointed out that this experimental testing with its singular individual statements cannot achieve the quality of results of an analytical investigation. All too often, however, precisely this analytical evaluation is also only feasible by means of simplifying abstractions. In these cases, forms of operational validation of the available simulations such as plausibility checks, sensitivity analyses or experimental parameter calibration through output comparison must be used.

### 3.4.4 Outlook

From today's perspective, the development of mechatronic design methods based on the V-model can therefore be easily understood in retrospect. Future development scenarios in the context of Industry 4.0 will also deal with the actual state of a production plant within its life cycle across domains in an increasingly integrated view from the perspectives already listed above:

(a) the entrepreneurial and business management perspective — Enterprise Resource Planning (ERP);
(b) the plant operation and technology view — Manufacturing Execution System (MES);
(c) the strategic maintenance perspective (condition monitoring and asset management).

One lesson that can already be learnt from the development of mechatronics is certainly that the specialist domains involved will not easily merge into a common, new discipline. The reason for this is that — as in the example of mechatronics — it is not only the interlinking of these specialist domains that is increasing but also their own degree of specialization. In order to cope with the growing complexity, the future will bring additional data and communication interfaces between the established tools in addition to cross-domain multifunctional tools — from standards to be developed across manufacturers to proprietary solutions from strong individual suppliers. Simulation technologies will also play an important role here, whereby the time axis for these observations will not depict the manufacture of a single product or the sequence along a single batch, but rather an entire annual or product cycle like time lapse. This means a qualitative simulation of plant life cycles, in whose compressed time sequences, fluctuations in quality and demand, as well as the aging and failure of production or data processing components, are reflected. Whether one should take the time-consuming route of working with a variation of a realistic selection from the explosively growing variety of individual events or whether a method will be established, which deals analytically with the probabilities of these events in a way similar to computational stability assessment of complex control loops, must remain unanswered at this point as a remaining subject of research.

The conclusion is by no means disappointing. The goals of technical development can already be outlined with sufficient clarity. Only in the coming years will the range of experience grow with an increasing number of application examples, as will the cross-domain or domain-linking functional scope of the software tools used.

As with mechatronics, we can expect to see exciting formats and concepts that bring together plant operations management, production technology and maintenance to create cross-domain solutions. The first examples have long since become reality. Raw parts communicate their individual further production recipe via RFID label to the system, a fault status of a component automatically logs off the associated machine with a repair forecast from the production system, the responsible maintenance mechanic triggers the ordering process for the spare part directly in the ERP system with the fault logging, etc. The prerequisite for implementing such links is, on the one hand, modeling of all the business processes involved, which usually also takes place across companies, and, on the other hand, an information architecture that is designed with foresight for such data-driven use cases.

# References

[1] DIN 61131 — 3:2014-06, Programmable logic controllers — Part 3: Programming languages (IEC 61131 — 3:2013); German version EN 61131 — 3:2013.

[2] Wintzenburg, J. B.: It doesn't fit. Airbus — The story of a Franco-German misunderstanding. Oder: Wie ein paar zu kurze Kabel einen ganzen Konzern in Schieflage bringen können, in: *Stern*, available online at: https://www.stern.de/wirtschaft/news/airbus-es-passt-nicht-3325122.html, last accessed: 25.07.2017.

[3] VDI/VDE Guideline 2206:2004, Development methodology for mechatronic systems, 2004.

[4] VDI/VDE Guideline 2206, Development of mechatronic and cyber-physical systems, 2021.

[5] Angermann, A., Beuschel, M., Rau, M. and Wohlfarth, V.: Matlab — Simulink — Stateflow — Grundlagen, Toolboxen, Beispiele, 10th ed., Oldenburg 2020.

[6] National Instruments: NI LabVIEW for Rapid Control Prototyping and Hardware-in-the-Loop Simulation, white paper dated 5 November 2013.

[7] Pauker, F.: OPC4Factory-OPC UA communication for manufacturing cells, https://publik.tuwien.ac.at/files/PubDat_252298.pdf.

[8] Schluse, M., Roßmann, J.: Von der Simulation zum Experimentierbaren Digitalen Zwilling und zurück. In Franke, J., Schuderer, P. (eds.): Simulation in Produktion und Logistik 2021, Cuvillier Verlag, Göttingen 2021.

[9] Stark, R., Anderl, R., Thoben, K.-D. and Wartzack, S.: WiGeP Position Paper: "Digital Twin", *Journal for Economic Factory Operation*, vol. 115, 47–50.

[10] Further development of the reference model for Industry 4.0 — component, structure of the asset administration shell, GMA/ZVEI status report, April 2016.

[11] Simulation and digital twin in the plant life cycle, VDI Status Report, February 2020.

# Chapter 4

# Rethinking Quality Assurance

Although high and consistent product quality has always played a key role in the production process, this issue takes on even greater significance in digital production. There are many reasons for this. One results from the trend for companies to strive for a high degree of adaptability (in response to market volatility), both at product and production system levels. This requirement cannot be met economically without integrated quality assurance in the production process before, during, and after each individual sub-process. It is extremely important to detect a defect as early as possible to save time, material, and ultimately costs by making appropriate corrections or even stopping the process. If, for example, a faulty print image is detected after printing page seven of a 50-page photo book, the page can be ejected immediately and the print run repeated. If a check or quality assurance is not carried out after each sub-process, e.g. at the end of the process chain, the entire book would have to be reprinted. In the worst case, only the customer notices the error. It is obvious that this is neither economical nor image-enhancing.

The same applies to the manufacture of the production systems themselves. The earlier a defect is detected, the sooner countermeasures can be taken or even discarded. Incidentally, this not only applies to actual production but also begins with engineering. If, as discussed in the previous chapter, continuous verification is carried out and documented during the development phase through extensive simulation and testing, it is much easier to correct undesirable developments and thus make a significant contribution to reducing the OEE.[1] It is important at this point to mention the possibilities of using simulation technologies and the digital twin, which can be used much more as alternative data sources for quality assurance in the future — even if this will not be the main subject of this chapter.

In this chapter, we want to look at the topic of quality assurance and the associated management from the perspective of Automation 4.0, illustrate various scenarios, and provide concrete examples. We introduce the most important topics before diving into the topic.

## 4.1  Overview of terms

The definition of quality according to the DIN EN ISO 9001 standard [1] considers the degree to which the intrinsic characteristics of an object fulfill the specified requirements. This object can be a service, a product, or software. There is a broad consensus that product quality has its origin in the quality of the manufacturing process. The concept of quality must also be applied to systems and ultimately to components in accordance with the principle of "total quality management". This leads to quality management in which systems are defined as a "group of interacting or mutually influencing elements" [1] which also includes the individual components.

The customer, the developers, and possibly other stakeholders, such as legislators, place various quality requirements on the end product during the development phase of automation solutions and thus ultimately on the automation system. In order to understand what exactly quality requirements are, we briefly define them as follows:

- Quality requirements are the totality of the requirements to be examined and specified for the quality of a unit, in each case with regard to the considered level of detail of the individual requirements.
- A unit can be both tangible and intangible, and in the case of a modularized automation system, the units under consideration are usually the components.
- Units, in turn, are subsequently represented by the description of characteristics[2] (characteristic properties) and their characteristic values (concrete characteristics), whereby it should be noted that different

---

[1]Overall Equipment Effectiveness (OEE) is a key figure that describes the productivity, profitability, and effectiveness of production plants/systems.

[2]In this section, characteristic and parameter are used synonymously.

standards may be applicable for different system components (e.g. hardware and software).

- It should be inherent in the characteristics and their characteristics (especially in automation technology) that they can be operationalized, and their concrete target values or target ranges can be specified.
- The characteristics of the areas must be measurable, i.e. an appropriate measurement technique must be defined for evaluating the quality parameters.

### ☑ Extract of relevant standards

The previously presented RAMI model and OPC-UA are among the most prominent Industry 4.0 standards:

- the IEC 62832 series "Digital Factory Framework",
- DIN NA 043-01-41 AA "Internet of Things (IoT) and Digital Twin",
- the IEC 63278 series "Asset Administration Shell for Industrial Applications",
- the ISO/IEC 20924:2021 standard,
- IEC 61360-4 — IEC/SC 3D — Common Data Dictionary,
- as well as the work of various working groups, such as,
- the ISO/IEC JTC 1/SC 41 Internet of Things and Digital Twin or,
- the IEC TC 65 "Industrial-process Measurement, Control and Automation".

*Quality management* is often defined as a strategic orientation for the production of quality, which is strongly correlated with the fulfillment of customer requirements.[3] A quality management system (QM system or QMS) is required to realize this, which is defined as follows in *DIN:ISO 9000*:

> *"A QM system comprises activities through which the organization identifies its objectives and determines the processes and resources necessary to achieve the desired results. The QMS directs and controls interacting processes and resources necessary to create value and realize results for relevant interested parties."* [1]

---

[3] A good overview of QM can be found in [8].

Among other things, a QMS includes the definition of quality objectives, the development of quality guidelines, the definition of procedures for planning, controlling, and monitoring quality and the establishment of measurement and evaluation systems. According to DIN ISO 9000:2015, the concept of *quality assurance* is part of quality management and aims to create confidence in the fulfillment of requirements. In the software sector in particular, quality assurance serves to ensure that the quality requirements of the software units are met.

## 4.2  What makes quality 4.0?

The question that arises is as follows: What influence do the developments of Industry 4.0 have on quality management, QM models, processes, and procedures? The answer is initially sobering: the literature shows that traditional quality management models are not aligned with the technological advances of digital production [2]. QM models are fundamentally based on the principles of economic efficiency, control, and systematic management — which is in no contrast to Industry 4.0. Nevertheless, with the fundamental process changes brought about by the development of Industry 4.0, many aspects of established QM models are becoming obsolete [2]. As a result, companies may consider the existing quality models to be irrelevant with regard to their specific context. In the worst-case scenario, the lack of coordination between QM development and the development of Industry 4.0 solutions could have a negative impact on effective implementation.

A key point is the technical focus of Industry 4.0, which changes the human roles in the system and makes it necessary to update the QM models. Industry 4.0 is data-based and creates new possibilities for data analysis, which previous QM models did not map well enough. Furthermore, the lean and agile structures of Industry 4.0 contrast with the traditionally extensive management systems in existing QM models. Finally, new collaborative business models require a transition in the QM models for managing the operations of networked companies. The focus here is on data analysis using big data analytics and AI. Traditional QM models need to be updated to address this new quality paradigm [2]. This change is also being driven by the massive introduction of sensor technology in production lines and other business processes, which can collect a wealth of data in real time — including and especially quality-related data. This data is stored, analyzed, and processed to identify patterns, make predictions, and support decisions [3].

Research and practice have adapted to this challenge: Parallel to Industry 4.0, the concept of *Quality 4.0* has developed, which primarily focuses on the new possibilities of using better and more efficient sensor technology, the larger database, and the possibilities of using machine learning for quality management as a whole. The technologies that are being introduced with Industry 4.0, be it IoT solutions or standards such as OPC-UA or the asset administration shell, are the basis for the introduction of new quality management using proven methods. All these mechanisms can and must be used to derive preventive quality assurance measures and predict quality in general (process and product). This approach is also referred to as data-driven quality management.

It should be noted that, in addition to the expansion of software aspects in automation (see Chapter 1), data and data processing also play a particularly important role, especially in the Quality 4.0 concept. As decisions, whether in the automation solution or fundamentally in the company, are increasingly based on data, it is important that this data is accurate, reliable, and relevant [4]. The quality of the end product, as well as the efficiency and effectiveness of automation solutions, are therefore directly dependent on the quality of the software and the underlying data. This is why ensuring data quality is an important pillar of QM, alongside the additional assurance of software and model quality [5]. The quality of data can be influenced by various factors, such as accuracy, completeness, consistency, timeliness, and relevance [6]. Ultimately, incorrect, or incomplete data can lead to incorrect decisions and significantly impair the effectiveness of automation solutions. It is therefore essential to implement suitable methods and technologies to ensure and improve data quality and to consider them as an inherent part of quality management in the development of automation solutions. Data governance structures, meta data management, or master data management systems, for example, are developed to ensure data quality and optimal use for QM [7]. These are also increasingly becoming a part of customers' corporate strategies [8].

Real-time data monitoring must be considered right from the start to make data-based quality assurance a success. This refers particularly to the continuous review of data as it is generated or captured — to ensure both data quality and end product quality. In this way, it is possible to gain immediate insights and react quickly to potential problems — this is already standard, especially in critical systems. It is common to set heuristic thresholds for critical characteristics (e.g. temperature) and model a response mechanism (e.g. shutdown and alarm). However, this also highlights the dangers — if this value depends on just one sensor, for example, and the data from this sensor is incorrect, enormous damage could be caused without there ever having been a real critical situation. The same applies to anomaly detection in real-time data monitoring. This aims to identify unusual patterns or deviations in the data that could indicate potential problems or errors. This method is particularly used in environments with high data throughput and in situations where quick reactions are required to prevent or resolve potential problems. To exemplify this, let's say a real-time data monitoring system is implemented in a manufacturing environment to continuously monitor the health and performance of its machines and processes. Anomaly detection helps to identify potential problems at an early stage and thus reduce unplanned downtime. Let's assume that a digital production facility has several production machines whose automation systems use a vibration sensor monitoring module, which is used at several locations. It continuously collects data and sends it to the production control system, where it is analyzed in real time to detect anomalies. This could result in the following scenarios:

- **Normal operation**
  In normal operation, the sensors indicate a constant vibration within an acceptable range.
- **Anomaly detection**
  A module signals an unusual increase in vibrations on a particular machine that goes beyond the normal range.
- **Correlation and validation**
  The data is compared with other data from the system and an automated check is carried out to determine whether the data is correct.
- **Data-based decision**
  If the data cannot be validated, the system sends information to the person responsible to inform them of the unusual data. If the data proves to be correct, the correlation can probably be used to react to sources of error (e.g. overheating) and possibly to automated troubleshooting (e.g. by lowering the room temperature or adjusting the production speed).

The new possibilities described for the use of additional sensors[4] in automated processes enable a new level of transparency and control.

---

[4]But it doesn't just have to be additional sensors. Measured values that are required to control the technological process anyway can also be used.

It becomes possible for each module to measure quality criteria or provide relevant data and in this way influence both the internal process and the overarching processes. Not only the control but also the derivation of actions (whether heuristic or AI-based) become more complex, depending on how many quality criteria and possibilities for action exist. The aim is to realize the shortest possible quality control loops[5] by measuring and feeding back quality data directly in the production line (in-line) [9].

Ultimately, Quality 4.0 is an evolution in quality management made possible by the technological advances and developments of Industry 4.0. By integrating sensor data and advanced analysis methods, companies can optimize their quality processes and ensure higher product and service quality.

### Summary & Outlook

As a result, a Quality 4.0 approach potentially enables companies to monitor every step of a production process in real time and to check and change process sequences according to the quality parameters. In the long term, quality criteria can also be linked to long-term use, for example. Quality 4.0 thus opens up the horizon not only for better control and prediction of product quality, but also for a faster response to potential problems or deviations. To achieve this, Quality 4.0 must be incorporated into the design of automation solutions [10].

## 4.3 Quality management and modularization

Modern quality management aims to understand quality as an integral part of all organizational processes. When developing automation solutions for a customer, it must be ensured that this approach can be supported. This idea is based on the assumption that the quality of an end product is linked to the quality of the internal, i.e. automated system processes. Essentially, the aim is to make the individual processes of the system describable and

---

[5]There are various interpretations of what exactly is meant by a quality control loop. However, reference is often made to the PDCA cycle (Plan-Do-Check-Act, also known as the Deming cycle). More generally, it is a cyclical sequence of activities in QM — including the example from the Six Sigma area in the following.

monitorable and to enable continuous improvement. These processes are ultimately to be understood as modules, as presented at the beginning. It should be remembered that an essential feature of modules is that each has input and output and therefore a defined beginning and end. In addition, modules should always be self-similar to allow them to be divided into further modules.

## 4.3.1 Measuring quality

The determination and measurement of process parameters remains an important part of quality management, especially quality control and quality assurance in automation technology, even in digital production. To ensure that both the process and the end product meet the specified quality standards and requirements, the various parameters, such as temperature, pressure, or speed, must be measured and analyzed. In this section, we take a closer look at determining and measuring the process parameters of the automation solution and focus less on the quality management of the development process.

The quality management of automation processes is based on the following:

- Careful identification of quality-related parameters (characteristics),
- determination of the measurement ranges and quantification of these (characteristic values),
- as well as correct measurement.

This applies both to the quality of the production processes themselves and to the product quality. Modularization makes it possible to systematically consider the individual modules with regard to the necessary quality and their parameters. It is also possible that there are process parameters that need to be recorded across all modules (e.g. for solutions that require certain ambient temperatures). In the course of modularization, it is also possible that complete modules are developed for quality assurance or quality management, e.g. camera-based systems for the optical control of process parameters with corresponding control for different quality characteristics.

In practice, there are various steps that lead to the goal. First of all, the relevant process parameters must be identified. There are two types in a modular system which can be distinguished:

- the module itself is regarded as a black box and the parameters are considered on the basis of the input/output or,
- the module is treated as a white box and the internal processes are modeled as process parameters (e.g. temperature monitoring).

In principle, the parameters to be used may vary depending on the type of production process or industry being considered. In general, however, the parameters should be able to adequately describe the production process or the process output. In particular, the process parameters should be determined in such a way that the influence on the quality of the end product is defined. In addition, it must be defined how the measurement is to be carried out within the process or within the module, e.g. which sensors are suitable. Various technologies are used to collect these parameters, which can vary greatly depending on the area of application, as the following examples show:

- **Sensors for measuring temperature, pressure, speed, fill level, or humidity**
  Temperature sensors are used in production systems for metals and plastics (e.g. melting furnaces or molding machines) to ensure that the materials are processed at the correct temperature. These sensors are placed directly in the system or individual components.

  In the electronics industry, humidity sensors are used to monitor that electronic components and devices are stored and processed at the correct humidity (the same applies in wood production, for example). These sensors are cross-modular and are usually placed in the environment — they should generally work autonomously.
- **Camera-based systems are used to monitor product quality, among other things**
  In standard scenarios, images of products are recorded and analyzed during the manufacturing process in order to detect defects and deviations. Such systems are used in the electronics industry, for example, to check the quality of components and assemblies. One example of this is the monitoring of soldered joints to ensure that all connections have been made correctly. Camera-based systems are also used in the automotive industry to detect cracks or deformations on component surfaces. This knowledge is applied to control the connection processes accordingly.
- **Analytical methods/analytical quality assurance**
  Analytical measurements are used, for example, in process engineering to determine the quality of raw materials or end products or to

monitor the quality of food products. For example, infrared analysis systems can be used to measure the fat content, moisture content, and other parameters of food samples. In the automotive industry, automated analytical quality assurance systems are used to monitor the quality of components and materials. For example, X-ray fluorescence analysis systems can be used to measure the composition of materials such as metals and plastics.

Once the relevant process parameters have been identified and the measuring instruments determined, it is necessary to determine the optimum values and corresponding tolerances for the parameters. The determination of these values can be achieved through statistical analyses of data from previous production runs, by applying industry-specific recommendations, or by carrying out test series or optimizations.

### 4.3.2  Analysis of quality

The actual measurement and monitoring require appropriate data collection and communication of the data in order to ultimately analyze it with the aim of identifying trends, patterns, and possible deviations from the previously defined parameters. Statistical methods and machine learning can be used to evaluate the performance of the production process and identify any weak points.[6] Examples of this are:

- **Multivariate regression analysis**
  Multivariate regression analysis makes it possible to investigate relationships between several independent variables (process parameters) and a dependent variable (quality characteristic). This method can be used to model the effects of various process parameters on quality in automated production processes. In metal processing, this method can be used to analyze the influence of process parameters such as temperature, pressure, and cooling rate on the quality of the manufactured parts, such as strength, hardness, or surface roughness. In [10], it is shown how these methods are used to predict the quality of wafers in production.
- **Neural networks**
  These are models inspired by the structure and function of the human brain. They consist of artificial neurons that are organized in layers

---

[6]An overview can also be found at [11].

(input layer, hidden layer, and output layer) and are connected to each other. Data is sent to the individual neurons. These decide on the basis of various weightings and functions (transmission and activation functions) which output signal should be generated and then pass this on to subsequent neurons. Neural networks learn, for example, by adjusting their internal parameters, the so-called weights, during training. The main purpose of neural networks is to recognize patterns in data and make predictions or decisions based on these patterns. They are particularly well suited to discovering complex, non-linear relationships in large amounts of data. For example, a network can be trained with existing data to minimize the error between the network's predictions and the actual data. They can also be used to predict quality characteristics based on process parameters in automated production systems — provided the database provides a sufficient basis for this. Examples from the metal industry show how convolutional neural networks can be used to automatically analyze surfaces, for example [12].

- **Anomaly detection**
  Examples are isolation forest or Support Vector Machines (SVM). SVM is part of supervised machine learning and is used to solve classification and regression problems. Isolation Forest, on the other hand, is one of the unsupervised forms of machine learning. Both can be used to identify unusual patterns or outliers in process parameters. This helps to detect potential quality problems in automated production processes at an early stage. An overview of the possibilities can be found, for example, in [13]. Furthermore, methods such as isolation forest are also used for intrusion detection systems (attack detection systems) for cyber-physical systems [14, 15].
- **Time series analysis**
  Time series analysis methods such as Autoregressive Integrated Moving Average (ARIMA) or Long Short-Term Memory (LSTM[7]) can be used to predict future values of process parameters in automated systems. This makes it possible to recognize quality problems in advance and initiate countermeasures (predictive quality). Applications of such models (LSTM) in the automotive industry show the potential. In one example, 94% of all process errors were predicted and it was possible to react before they occurred [11].

---

[7]LSTM is a very special type of neural network.

### 4.3.3 In detail: Digital image processing as a quality assurance process

Digital image processing enables fast, automated, and precise inspection of products. In fact, according to [16] around 80% of all digital image processing applications are in this area. The importance of digital image processing has increased enormously in recent years. In the medical field, e.g. dermatology, some AI systems are already better than experts according to [3] ("Deep learning outperforms 136 out of 157 dermatologists").

One example of the use of digital image processing for quality assurance is the inspection of printed circuit boards in the electronics industry.

Printed circuit boards are an essential component of many electronic devices and must be checked for defects during the manufacturing process to ensure high product quality and functionality. Defects can include missing or incorrectly positioned components, solder bridges, or short circuits. Automated Optical Inspection (AOI) using image processing has established itself as an effective method for quality control of printed circuit boards. AOI systems use cameras and lighting equipment to capture high-resolution images of the PCBs. These images are then analyzed using image processing algorithms to identify defects or irregularities. Various image processing techniques such as thresholding, morphology, edge detection, and pattern recognition are used to extract the characteristics of the PCBs and assess the quality of the components and solder joints.

A concrete example from practice [17]: An automated optical inspection system can be used as part of the post-reflow process, directly after the reflow soldering process in the reflow oven.[8] In an imaging AOI sub-process, optical images of the region of interest (ROI) can be captured by a moving multiple-camera system. These can be set depending on the placement plan and the properties of a product. Multiple cameras enable efficient in-line integration without time delays. The use of powerful software components and solutions, such as modern frame grabbers and intelligent control software, in combination with precise electromechanical components (e.g. synchronous linear motors), enables high inspection speeds. In the next step, this information must be transferred to the image processing system. Depending on the meta information of the image section to be inspected, the corresponding

---

[8]Examples on this topic can be found in [17].

product components or solder joints are assigned to the analysis algorithms. These algorithms are capable of detecting a large number of soldering, assembly, and printing defects in accordance with the acceptance criteria of the applicable industry standard, whereby these algorithms can be designed very differently [17].

Image processing for quality assurance is not only widespread in the semiconductor industry but has found its way into many areas.

The use of computer vision and machine learning technologies can automate defect detection during the quality control process, for example in the printing industry, and thus increase efficiency. In many cases, it can be shown that there is a reduction in the costs associated with quality control and increased defect detection accuracy. To make this more concrete, a further example from the literature is taken up.

In the example shown in [16], a deep neural network for industrial quality control in the printing industry is implemented in the following way:

First, experimental setup for training and testing the deep neural network includes providing the necessary hardware and software. This may include the selection and setup of powerful computers, specialized graphics cards for machine learning (e.g. graphics processors and GPUs), as well as the installation and configuration of the required software frameworks and libraries such as TensorFlow or PyTorch. In addition, specific devices for data acquisition, such as cameras, must be installed and the corresponding data collected. Data is collected via a central data storage system, whereby the configuration of the hardware modules is not described further at this point (e.g. communication, etc.). The decisive factor here is that data processing should also be divided into different modules. First, the data must be preprocessed. These pre-processing steps serve to standardize and organize the input data for the Deep Neural Network (DNN) in order to make the training and evaluation of the DNN more efficient. Various steps may be required here, e.g.,

- **Noise reduction**
  Digital images can be distorted by various factors, such as sensor noise or environmental conditions. Filters, such as the Gaussian filter or the median filter, are used to reduce noise and improve image quality.
- **Histogram equalization**
  Intensity values of the image can be redistributed to improve contrasts. This can be particularly useful if the image was taken under poor lighting conditions. For example, brightness adjustment can be implemented

by histogram stretching to ensure consistent brightness across all cylinder images [18].

- **Segmentation**
  This involves dividing the image into different segments or regions in order to emphasize certain objects or features in the image. Methods such as thresholding, edge detection, or watershed segmentation can be used.
- **Morphological operations**
  These operations, such as erosion and dilation, are used to improve the structure or shape of objects in the image or to remove unwanted pixels.
- **Scaling and resizing**
  In some applications, it is necessary to scale or reduce the image to a certain size to increase processing speed or to adapt it to certain algorithms and convolution windows. This can be supported by prior analysis.
- **Color space conversion**
  Images are often converted from color to grayscale images in order to reduce complexity. In certain applications, conversion to other color spaces such as HSV or LAB can also be useful.
- **Edge detection**
  Edges in the image often represent important information. Algorithms such as the Sobel or Canny operator can be used to highlight these edges.
- **Calibration**
  In industrial applications, especially in robotics or measurement applications, it is often necessary to calibrate the camera in order to correct distortions and perform accurate spatial measurements.

However, the most important step is an annotation, especially for guided learning methods. The annotation can be binary or much more complex, depending on the particular case. Automatic selection and labeling of the data is also possible, such as [18] shows. The next step is to determine the basic architecture of the CNN based on the data. In principle, CNNs consist of different layers:

- **Convolutional**
  These layers use predefined filters to extract features from the input data.

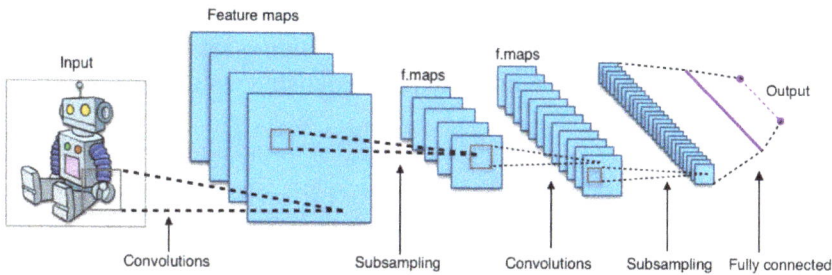

**Figure 4.1.**   Typical CNN.

*Source*: Aphex34, CC BY-SA 4.0 via Wikimedia Commons https://commons.wikimedia.org/wiki/
File:Typical_cnn.png.

- **Pooling**
  They are used to reduce data complexity by dividing the data into different sections or areas and reducing dimensionality.
- **Fully connected**
  These form additional neuronal connections between the layers. This enables the network to recognize complex relationships between features and make sophisticated predictions.

These layers are joined together as shown in Figure 4.1.

The decisions about the exact architecture and composition are very complex and cannot be described here. Once the architecture has been determined, the next step can be taken: *validation*.

For this purpose, the data must be divided into training, validation, and test sets. A common division is 70% training, 15% validation, and 15% test. It should be ensured that all classes are evenly distributed across the sets. Cross-validation is often used to increase the informative value and to assess the generalization capability of a model. In cross-validation, the dataset is divided into several subsets, and the model is trained and validated several times, with different subsets being used as training and validation data each time.

- **K-Fold**
  Data is divided into k approximately equal parts. Each subset is used once as a validation dataset, while the others are used for training. This is repeated *k* times and the mean error over all runs is calculated. This method is widely used but can be time-consuming.

- **Holdout**
  The data is split once into two parts for training and validation. This speeds up the execution but may be less reliable with small data sets.
- **Leave-One-Out**
  A special form of K-Fold where k is equal to the number of observations. Each observation is used once as a test data set. Also known as leave-one-out cross-validation (LOOCV).
- **Repeated random subsamples**
  Uses the Monte Carlo method to create multiple random splits into training and test data. The results are summarized, and data points can be used multiple times for testing.
- **Re-substitution**
  Uses all data for training and assessment. This method can lead to over-optimistic performance estimates and should be avoided if sufficient data is available.

Each of these techniques has its own advantages and disadvantages, which should be taken into account depending on the size of the data set and the specific application context. Validation represents quality assurance. Once the model has been developed, it can be integrated into a production environment. In the following section, we return to Quality 4.0 and its practical application.

## 4.4 Bringing Quality 4.0 into the application

As an example of the application, we assume that the adhesive binder developed in Chapter 2 for integrated quality assurance follows a lean management approach (Six Sigma). Six Sigma was originally developed by Motorola in the 1980s and is widely used. The name can be derived as follows:

A process that is within +/-6s, or Six Sigma, of the center line of a control chart is considered to be well regulated. Sigma refers to the standard deviation and is often represented by a small "s" or the Greek symbol "$\sigma$". This means that there are only minimal deviations from the standard and the process performance remains within the specified tolerance limits. Six Sigma is also often represented as $6\sigma$. The main goal of Six Sigma is to minimize process variation and maximize the quality of results by focusing on customer requirements and making data-driven decisions [19].

In automation, Six Sigma plays a crucial role in quality control. It uses statistical analysis to identify the causes of errors or deviations in an automation process and then take action to correct these errors. One goal is to reduce the number of defects per million opportunities (DPMO) and thus achieve a process that is virtually defect-free. Six Sigma uses the Define, Measure, Analyze, Improve, Control (DMAIC) framework to minimize process variability and maximize quality [20].

The scenario presented shows the implementation of a camera-based quality inspection module in our modular perfect binder to improve spine processing. The implementation of this project can be well supported by DMAIC.[9] A possible sequence/section of the DMAIC procedure for the given scenario is as follows:

## 4.4.1 Define

In the Define phase of the DMAIC cycle, the project objectives and scope are determined, and the business process/case is developed. Normally, the first step is to define the project, i.e. select the project based on business objectives, and define the project goals and scope. A team is then formed, selecting team members with the required skills, and developing a project plan. This is accompanied by the creation of a schedule and the identification of required resources. At the same time, customer requirements are determined, i.e. a detailed identification of customer needs and expectations is carried out on the basis of which the business process/case is developed.

The aim of the exemplary project is to improve the back processing, as inconsistencies were previously identified in this area. This manifested itself in unevenly milled specimens or inadequate preparation for adhesive application and was identified as a Critical To Customer (CTC) attribute or requirement. This requirement can be determined more precisely by identifying quality parameters. In this case, these are the parameters for the movement of the milling cutter, more precisely the depth Z, as well as the widths a1 and a2. Furthermore, the surface quality was selected as a parameter, as a correlation between future adhesive bonding and the surface structure was assumed. The automation system in its present form only has a small amount of process data. This is to be changed by integrating a camera-based quality inspection. The sub-goal is to detect errors in the production process at an early stage and rectify or control them.

---

[9]This example serves only as an illustration and cannot represent the complexity of the entire process in full detail.

### 4.4.2 Measure

In the Measure phase of the DMAIC cycle, important aspects of the current process or modules are measured and collected. This means that measurement parameters are identified and defined (e.g. KPIs[10]). In addition, current process data is collected, recorded, and documented. As a rule, this also includes the creation of a detailed process flow diagram. To ensure the reliability and accuracy of the collected data, a baseline for process performance is established by analyzing the collected data and validating the measurement systems. This phase provides a clear understanding of the current process performance and lays the foundation for the subsequent analysis and improvement phases.

In this case, we start by collecting data on the current performance of back processing. For this purpose, the modules used to date are examined more closely and possibilities for data extraction are explored. So far, there are only a few data points apart from the output, so random samples are collected. The measurement concerns the current depth of the milling machine, the widths a1 and a2, as well as the surface quality across various modules used by different customers, as well as an in-house test position with camera (see Figure 4.2). The results are also measured. At the end of the process, the measuring system is validated to ensure the accuracy and reliability of the measurements.

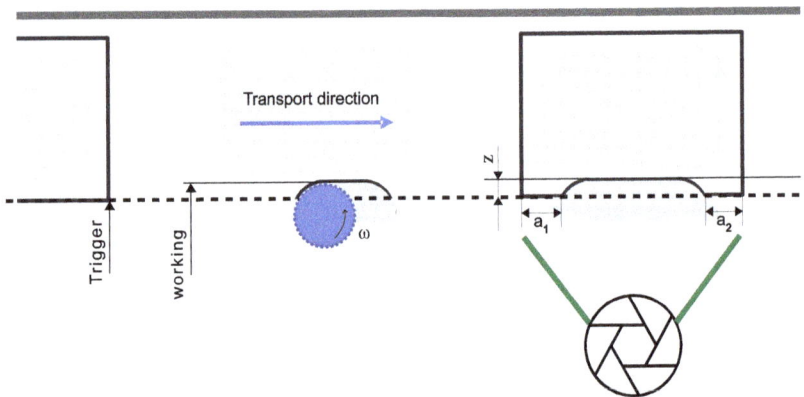

**Figure 4.2.**    Internal camera after milling process in a perfect binder.

---

[10]Key performance indicators, or KPIs, are measurable values that indicate the effectiveness of companies, departments, or projects in terms of achieving important goals and performance in certain key areas.

### 4.4.3 Analyze

The analysis phase in the Six Sigma DMAIC cycle is important for identifying the main causes of problems or defects in a process. The basis for this is the data analysis of the data collected in the Measure phase. In practice, this means using statistical analysis and presentation methods such as histograms, Pareto analyses, correlation, and regression analyses. The second important point is a process analysis, the review of prepared process flow diagrams and other schematic representations as well as the identification of bottlenecks, variations, and other process weaknesses. This goes hand in hand with a root cause analysis. Various cause-and-effect diagrams (Ishikawa or fishbone diagrams), the 5-Why method, and other root cause analysis techniques are used for this purpose. These root causes must be validated and ultimately prioritized, i.e. the identified root causes are evaluated and prioritized based on their impact on quality or performance.

In the scenario presented, the data collected is analyzed to identify the causes of the inconsistency in back machining. The collected data shows that there are significant differences in the milling depth. An error in the placement of the backs means that in some cases it is not optimally inserted. It is also found that the rotary cutter needs to be recalibrated after a certain number of book blocks with different materials. In addition, signs of wear and tear appear on the surfaces of the book spines.

This is followed by an evaluation of various technologies and approaches for improving spine processing or for better scheduling of maintenance work (e.g. replacement of the milling head) or the repositioning of books. The criteria presented in Chapter 2 must be included in the evaluation of the various technologies and approaches for improvement.

Modularization now allows various scenarios for the integration of quality assurance components. Depending on the error source analysis, it would also be possible at this point to select sensor technology other than camera-based systems and thus deviate from the original objectives. For example, sensors could be used that can determine the movements of the milling head and the position of the book (distance measurement, etc.). At the same time, the importance of the data situation, the underlying sensor technology, its integration, quality, and access for quality management in general becomes clear. Without detailed and high-quality data, however, the work in the first two phases is sometimes very time-consuming, as the data must first be recorded, experiments carried out or even digitized.

In the scenario presented, it is now also determined that measuring the depth of the cutter is not sufficient and the focus is on a camera-based

**Block back milling**

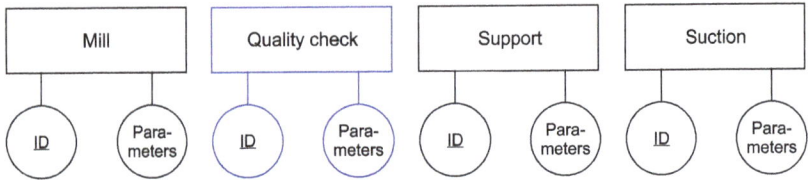

**Figure 4.3.**    Extension of QA as an internal function.

system. The reason for this is that the surface structure is a quality attribute, and this can be captured well with camera systems, and other methods are also not profitable, for example. However, there are different approaches as to how the camera system should be integrated into the existing system. For example, a measuring device (camera) could be integrated into the existing "Edit back" module (scenario 1). The extension of the module is shown in Figure 4.3 and relates to a further internal function (as presented in Chapter 2). For this purpose, the module is redesigned as an additional derivative — including the possible states and interfaces.

Another implementation scenario would be for extra camera modules to be developed and introduced into the process as additional modules (scenario 2). The additional benefit is easy to see. It would allow old automation systems to be retrofitted with very little effort. At the same time, such a camera module could also look the same in different places (reuse). The differences in implementation would then be reflected in the configuration of the detection and analysis of quality criteria. In other words, in the integration of suitable software modules and the realization of control elements and not in the design of the hardware components. In both scenarios, the aim is to detect defects or quality deficiencies immediately or at an early stage and to adapt the process accordingly. The scenarios differ in terms of the effort involved and the subsequent use of the development. At the same time, it should be noted that an attempt should always be made to strive for several measuring points with different methods for one feature in order to guarantee the accuracy of the data (as already described above). In the present case, for example, control data from the milling machine could be combined with the camera data.

At the end of the process, a suitable camera system is selected and implemented to record the relevant quality parameters. In this case, scenario 2 is chosen (see Figure 4.4).

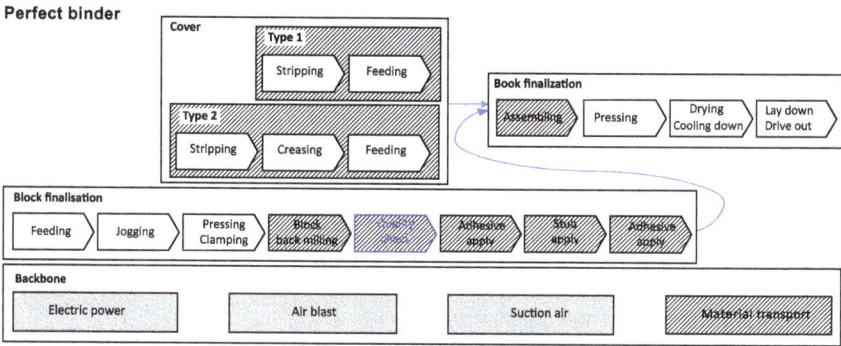

**Figure 4.4.** Extra module for video-based quality control of back processing (scenario 2).

## 4.4.4 Improve

In this phase of the DMAIC cycle, solutions are developed, implemented, and tested based on the findings of the previous analysis phase. The transitions between the phases are handled smoothly. Sometimes a technology assessment is already carried out in phase 3 or the focus there is exclusively on root cause analysis. However, the final solutions take shape in the Improve phase, where various creative methods are used to find solutions. It is particularly important to select the most effective and efficient solutions. One of these methods is the Design of Experiments (DoE). This is a statistical method for understanding and modeling the effects of different factors on a response variable. This is achieved by systematically varying the input factors and observing the effects on the response variable, thereby identifying the relationships between the factors and the optimal setting of the factors for desired outcomes.

Selected solutions are then transferred to pilot projects, for example, as a minimum viable product (MVP) on a smaller, controlled scale for validation. Modularization supports simplified implementation. The aim is to measure and check the process performance after implementation of the solutions in order to assess their effectiveness. Validation is the final step before the solutions are fully implemented. The Improve phase is therefore crucial for achieving significant improvements and the set goals.

A practical example of the application of these principles is the implementation of a camera-based quality inspection module in spine processing. This module, based on decisions made in the previous phase,

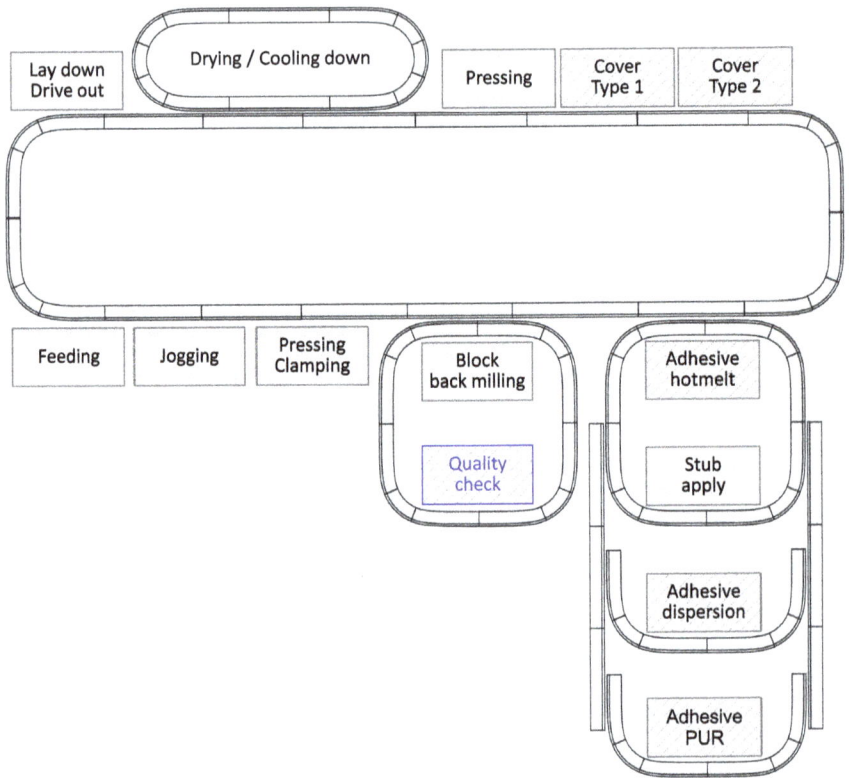

**Figure 4.5.**   Module diagram for an adhesive binder according to case study 2 with the extension of a test module.

measures the depth, width, and evenness of the milling of each spine in real time. It uses image processing algorithms to identify potential defects, such as correct depth, surface finish, and spacing. By integrating feedback loops, milling processes can be repeated if, for example, too little material has been removed. The transport system is also adapted so that faulty products can be ejected automatically. The module diagram for this case is outlined in Figure 4.5.

### 4.4.5  Control

In the control phase of the DMAIC cycle, the improvements achieved are maintained through regular monitoring and control of the optimized process or module. This includes reviewing process performance to identify and correct deviations at an early stage. Adjustments can be made based on the collected data and analysis results, such as readjusting the milling machine or modifying the milling speed to ensure continuous optimization in the respective application areas.

The successful implementation of a project is usually accompanied by training and an informed handover to customers. It is also the basis for a thorough reflection on the success/failure of the project and the identification of further potential for improvement. This can lead to the development of additional services, such as predictive maintenance, and the evaluation of further automation options, always taking economic efficiency and technical feasibility into account.

The implementation of Six Sigma and specifically the DMAIC cycle in the context of the modular perfect binder presented illustrates the potential offered by a structured approach to quality assurance and process optimization in terms of Automation 4.0. From precise definition and accurate measurement to detailed analysis, the project was able to identify and address significant weaknesses. The successful application also shows that a significant increase in quality can be achieved through a consistent focus on customer requirements and the use of data-driven decisions. Thanks to the modularity of the system and the flexibility of the DMAIC approach, specific quality challenges can be addressed in a targeted manner and innovative solutions can be implemented.

# References

[1]  DIN EN ISO 9000:2015-11: Quality management systems — Fundamentals and terminology (ISO_9000:2015); German and English version EN_ ISO_9000:2015.

[2]  Asif, M: Are QM models aligned with Industry 4.0? A perspective on current practices. *Journal of Cleaner Production,* vol. 258, p. 120820, 2020. https://doi.org/10.1016/j.jclepro.2020.120820.

[3]  Schuh, G., Anderl, R., Gausemeier, J., *et al.:* Industry 4.0 maturity index: Shaping the digital transformation of companies. Herbert Utz Publishing House 2017.

[4]  Wang, R.Y. and Strong, D.M.: Beyond accuracy: What data quality means to data consumers. *Journal of Management Information Systems*, vol. 12, pp. 5–33, 1996. https://doi.org/10.1080/07421222.1996.11518099.

[5]   Zonnenshain, A. and Kenett, R.S.: Quality 4.0-the challenging future of quality engineering. *Quality Engineering*, vol. 32, pp. 614–626, 2020. https://doi.org/10.1080/08982112.2019.1706744.

[6]   Batini, C. and Scannapieco, M.: *Data and Information Quality*. Springer International Publishing, Cham 2016.

[7]   Schumacher, J. and Weiß, P.: Process and data governance as a strategic approach to improving process and data quality in companies. *HMD*, vol. 48, pp. 82–89, 2011. https://doi.org/10.1007/BF03340590.

[8]   Werner, T.: How companies use data — and what slows them down, 2022, https://www.dihk.de/de/themen-und-positionen/wirtschaft-digital/digital-isierung/digitaler-aufbruch-mit-hindernissen/wie-betriebe-daten-nutzen-und-was-sie-dabei-ausbremst-66654, accessed 01 November 2023.

[9]   Lanza, G., Haefner, B., Schild, L. *et al.*: In-line measurement technology and quality control. In: Gao, W. (ed.) *Metrology*. Springer Singapore, Singapore 2019, pp. 399–433.

[10]  Sader, S., Husti, I., and Daroczi, M.: A review of quality 4.0: Definitions, features, technologies, applications, and challenges. *Total Quality Management & Business Excellence*, vol. 33, pp. 1164–1182, 2022. https://doi.org/10.1080/14783363.2021.1944082.

[11]  Tercan, H. and Meisen, T.: Machine learning and deep learning based predictive quality in manufacturing: A systematic review. *Journal of Intelligent Manufacturing*, vol. 33, pp. 1879–1905, 2022. https://doi.org/10.1007/s10845-022-01963-8.

[12]  Liu, Z., Tang, R., Duan, G., *et al.*: TruingDet: Towards high-quality visual automatic defect inspection for mental surface. *Optics and Lasers in Engineering*, vol. 138, p. 106423, 2021. https://www.sciencedirect.com/science/article/pii/S0143816620312124.

[13]  Rostami, H., Dantan, J.-Y., and Homri, L.: Review of data mining applications for quality assessment in manufacturing industry: Support vector machines. *International Journal of Metrology and Quality Engineering*, vol. 6, p. 401, 2015. https://doi.org/10.1051/ijmqe/2015023.

[14]  Laskar, M.T.R., Huang, J.X., Smetana, V., *et al.*: Extending isolation forest for anomaly detection in big data via K-means. *ACM Transactions on Cyber-Physical Systems*, vol. 5, pp. 1–26, 2021. https://doi.org/10.1145/3460976.

[15]  Humayed, A., Lin, J., Li, F., *et al.*: Cyber-physical systems security-a survey. *IEEE Internet of Things Journal*, vol. 4, pp. 1802–1831, 2017. https://doi.org/10.1109/JIOT.2017.2703172.

[16]  Linß, G.: Qualitätsmanagement für Ingenieure, 4th, updated and expanded edition. Hanser, Munich 2018.

[17] Matthies, P.C.: Integration of intelligent classifiers and inspection methodologies in the automatic optical inspection of automotive products in SMT production, University Library of Clausthal University of Technology 2023.

[18] Villalba-Diez, J., Schmidt, D., Gevers, R., *et al.*: Deep learning for industrial computer vision quality control in the printing Industry 4.0. *Sensors* (Basel) 19, 2019. https://doi.org/10.3390/s19183987.

[19] Harry, M.J. and Schroeder, R.: *Six Sigma: The Breakthrough Management Strategy Revolutionizing the World's Top Corporations*, 1st edn. Currency, New York 2000.

[20] Pande, P.S., Neuman, R.P., and Cavanaugh, R.R.: *The Six Sigma Way: How to Maximize the Impact of Your Change and Improvement Efforts*, 2nd edn. [rev. and fully updated]. McGraw-Hill Education, New York 2014.

# Chapter 5

# Modular Automation in Practice

The previous chapters presented the modular, function-, and object-oriented design of individual machines and systems in theory and used practical examples as a solution for increasing efficiency throughout the entire life cycle. It was shown that integrable and autonomous modules in the form of mechatronic components require different prerequisites for optimum operation and that automation-related boundary conditions must also be taken into account. In this chapter, we look at the individual aspects that need to be considered in a modular structure from the perspective of practical and automation technology.

We address the topics of step-by-step modularization of series machines at the beginning as well as the selection of automation systems including communication and safety technology. We also address the question of how real-time capability can be maintained or even improved in a modular system. Problem areas are identified, and possible solutions are pointed out so that the development of a modular production system will become a successful project, whether that is in a step-by-step process or starting with an idea.

## 5.1 Successive modularization

Series machine construction, as it is currently practiced, is as multifaceted as there are manufacturers of machines and systems. Hardly any manufacturer pursues exactly the same strategy to adapt their products to the individual needs of their customers. This is fueled even further by the fact that suppliers also pursue their own modularization strategies, regardless of whether they are dealing with automation technology components or complete mechatronic units, such as pumps, laser sources, or hydraulic units. Consequently, there is a heterogeneous landscape of different modular solutions and strategies.

The previous chapters describe a modularization process from the beginning and show how an almost ideal and, above all, sustainable modular structure can be achieved. In practice, however, it is rare for the development of a modular production system to begin on a blank sheet of paper. Rather, non-structured or partially structured systems are only subjected to modularization when the complexity becomes unmanageable, or the constraints listed in Chapter 1 become apparent.[1] If the decision is then made not to completely redesign the system but only to partially modernize it, there is a great opportunity for an initial modular design without completely questioning the structure of the existing system. We want to use the term *successive modularization*, which could also be called brown-field modularization and means the creation and implementation of modules in existing systems.

It starts when, for example, a module for a new technological requirement is to be developed by the machine manufacturer itself or purchased from a supplier and integrated into an existing machine. Another reason could be the lack of time or financial resources for a completely new development. Successive modularization spreads the initial time and financial outlay over several projects and is therefore easier to manage economically, although the sum of all individual efforts is unlikely to be less than a thorough modularization from scratch.

There are also several technical obstacles. If, for example, the gluing station in a non-modular designed perfect binder is to be designed as an independent mechatronic unit in the future, regardless of whether as an integrable or autonomous module, some additional design constraints must first be analyzed. It must be clarified how one can detach the unit mechanically and electrically from the overall design and add it back on in a new design form. The drive concept of the module must also be redesigned if, for example, the drive tasks required for the adhesive application were previously only realized via mechanical couplings, such as belts or bevel gears. Another problem arises if there should not be any

---

[1]The same problem is known from plant construction. The terms "green field" have become established for the construction of completely new plants and "brown field" for the modernization of existing plants. These terms are also used in software development when newly developed parts are to be integrated into existing projects.

mechanical energy transmission from the rest of the machine any longer and servo drives are now to be used because what the mechanics previously had to perform is now added as a dynamic requirement for the automation technology.

It must now be ensured that the required drive technology can also be decentralized and operated with a suitable field bus so that no qualitative restrictions arise due to insufficient synchronization. Both the field bus and the rest of the control technology must have the required *real-time capability* in order to achieve this.

The same applies to the safety technology. If previously only the main drive was shut down during an emergency shutdown, the decentralized drives and the decentralized peripherals must now also be included in the safety concept. If additional benefits are to be generated with intelligent safety solutions, e.g. for setup and retooling, then it must be clarified how these safety functions can be implemented at modular level and in the context of the overall system.

Although these considerations are part of every modularization process, they are particularly important in a migration process. The problem here is that it is probably a well-functioning machine that is now being interfered with quite deeply and that should improve in as many aspects as possible in the end. This includes not only proven mechanics but also stable and reliable suppliers, an automation system that is assumed to know everything, as well as the specific application, which involves many man-years of software engineering. In addition, there might be many systems and customers worldwide who rely on proven technology and have spare parts in stock. Therefore, it is not an easy decision whether to modularize each type of adhesive at least partially or to design a modified complete machine.

Even if we already know the answer, it may look very different in certain circumstances. Let's just imagine that the automation system is not suitable for synchronizing decentralized drive technology with the required accuracy or that a safely reduced speed required by the safety concept is not supported. Furthermore, the integration environment[2] must be adapted, which often leads to structures which are even more confusing and prone to errors. These are just some of the obstacles that might be in the way of using at least partial modularization. What should you do to continue modularizing successfully?

---

[2]This refers to the rest of the machine, for which we introduced the term *backbone* in Chapter 2.

We want to answer this question again from the point of view of automation technology and assume that all technological, mechanical, and electrical boundary conditions have been positively clarified. I4.0 conformity should not play a role at first, but we want to get as close as possible to a suitable function- and object-oriented solution to be able to take this requirement into account later.

### 5.1.1 Scenarios of a successive modularization

The problems and approaches to solutions for the typical scenarios in practice are again demonstrated using the example of a *perfect binder*. For this purpose, we choose an initial situation that can, in principle, also apply to processing machines in general.

Let's further assume that the automation concept of the non-modularized machine is structured according to Figure 5.1 and has the most important properties listed in the following:

- All control tasks are processed quickly and accurately enough in a CPU.
- The input and output modules are located directly on the backplane of the control unit.
- The drive technology is available for the realization of electronic gears (e.g. for the function *Apply adhesive*) via one or more dedicated controllers.
- The position of the main drive is transmitted synchronously to the CPU and the axis controllers via the field bus.
- Some inputs and outputs must be controlled very precisely depending on the position of the main drive.
- The safety technology is integrated into the control system and should have safe motion functionality.
- The required reaction time for control and drive technology is < 1 ms.[3]

---

[3] This value corresponds to the requirements of a medium performance class perfect binder. Modern plastics processing, printing, and packaging machines also have requirements in the order of 400 $\mu$s and shorter.

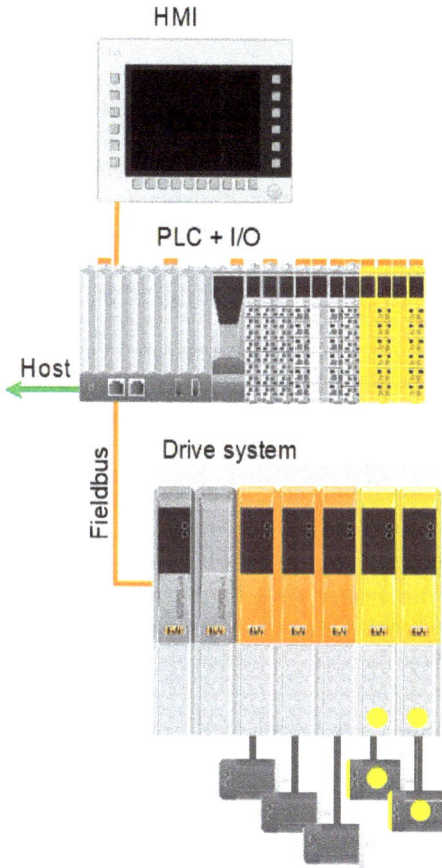

**Figure 5.1.** Initial configuration of the automation of a non-modularized machine.

- The entire operation and visualization are implemented via a main control panel in combination with an industrial PC.

If a technological function (e.g. the *application of adhesive* in the form of a *gluing station*) is to be detached from this configuration, the following scenarios are conceivable in principle, which we want to demonstrate with the knowledge of the design and function of integrable and autonomous modules with the corresponding boundary conditions.

## 5.1.2  Decentralized hardware is possible

**❗ Delimitation**

The utilized automation technology supports the decentralization of drive systems with sufficient accuracy for the technological functions and can also operate decentralized inputs and outputs with the required response time.

In this case, the first *option* is to create an *integrable module* (Figure 5.2). The input and output modules required for the gluing station are provided with a bus coupler and coupled to the extended field bus. The same is done with the required drives and an optional HMI device. The devices can be integrated directly into the mechanical system without an additional control cabinet or housed in a module's own control box, provided the required components are available with the appropriate degree of protection and are functionally compatible.

The following boundary conditions must be observed for this variant:

**Compatibility**
The input and output modules must be functionally compatible, i.e. it should not matter for the function whether the input and output channels are controlled directly by a control CPU or via a field bus and bus coupler. It is irrelevant whether they have the same design, as it would be the case for modules with protection classes IP20 or IP67.[4] The same applies to drive technology and HMI devices, especially if these components do not have their own intelligence.

**Real-time capability**
Decentralization via a field bus inevitably causes a change in the reaction time (Section 5.2.1). In the configuration shown for the operation of the inputs and outputs, this means that the resulting cycle time must still be

---

[4]The IPxy protection class describes the degree of protection against contact, foreign bodies (first digit), and moisture (second digit) in accordance with DIN EN 60529. IP67 means that an electrical device is dust-tight (6) and protected against temporary immersion in water (7). In contrast, an IP20 device is only protected against the ingress of solid foreign bodies with a diameter $\geq 12.5$ mm (2) but not against the ingress of moisture (0).

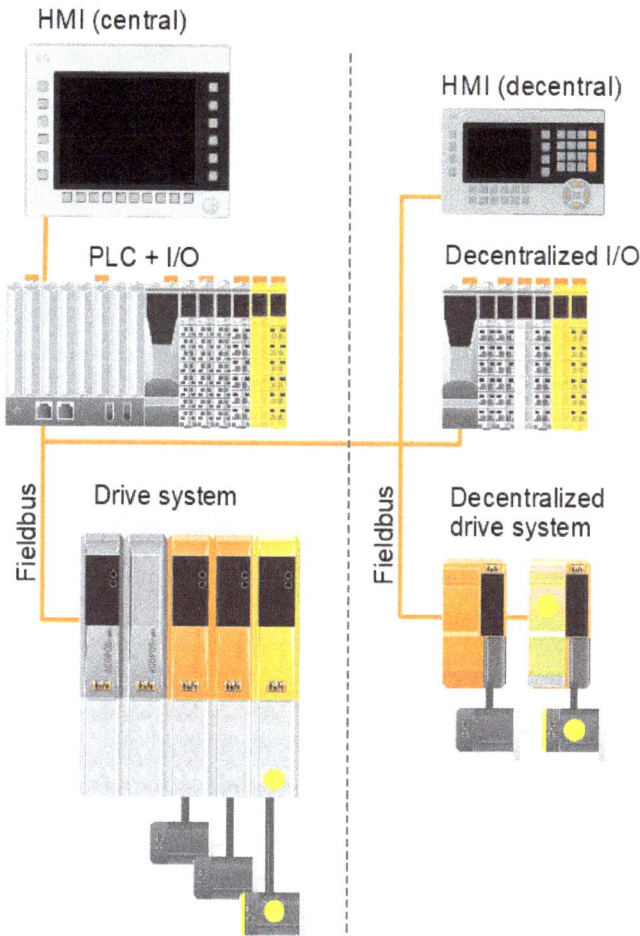

**Figure 5.2.**   Automation according to case 1 with an integrable module.

sufficiently short for the technological function to be fulfilled. The same applies to the drive technology, even if it is already coupled via the field bus in the output configuration. However, if other designs of servo drives or frequency inverters are used, this can lead to a change in behavior.

**Safety technology**

With the partial placement of the safety modules away from the central controller and toward the decentralized bus node, it must be ensured that no fundamental change is made to the safety configuration, as it would be the case by adding additional safe sensors or actuators, for example (Section 5.3.3). In this case, this automatically represents a change to the safety configuration and must be evaluated in accordance with Annex 4 of the Machinery Directive DIN EN ISO 13849 [1]. As a result, it may be necessary for the safety of the machine to be reassessed either by an accredited testing institute or by the manufacturer itself and documented in the so-called *manufacturer's declaration*. However, even if the basic safety configuration is retained, the safety concept of the automation system must also support a new placement of safety components without program changes. If the integrated safety technology shown in Figure 5.2 is not possible with the system used, a separate safety bus or, as a last resort, a safety bus connected to the system would also be possible. Additional parallel wiring of the safety-relevant signals is possible with a field bus. However, these are unpalatable alternatives, as they counteract the effects expected from modularization because they cause additional effort and increase complexity again.

**HMI components**

If a separate control point is provided for the new module to be created, two further problem areas arise. The first is the hardware integration, which is carried out directly via the common field bus. In addition to a suitable communication protocol, this coupling also requires the bus to be able to transport asynchronous data without affecting the transmission of synchronous data traffic (Section 5.4.1). This is not possible with all established field bus types, at least not if the available bandwidth is low and/or there are high demands on the accuracy and speed of the synchronous data traffic.

On the other hand, there are the software options, which in turn depend on the selected hardware equipment of the HMI panel. For example, if the panel has its own or is combined with additional intelligence (e.g. an IPC), it could also be programmed in its own and/or separate visualization project. In this case, the load on the field bus would also be the lowest, as only the qualitative data is transmitted, and it would be an important step toward reducing complexity.

If, on the other hand, a panel with web-based visualization is used that is operated via a web server running in the central CPU, the intelligent main operating panel, or an IPC, then this is certainly the best choice on the engineering side, but there may be higher requirements for asynchronous data traffic. After all, the operator does not have time to wait a long time for a requested frame change just because a fast-running synchronous drive calls up synchronous position data in ms or μs cycles and thus utilizes the field bus. In this case too, a separate bus connection would be possible but also the worst choice.

### Field bus
In addition to the requirements mentioned above, there is also the capability of a (hot) plug-in. For example, if the glue station is to be offered in various derivatives, then a different hardware configuration of the module is to be expected. If the gluing station is converted, these changes must be recognized automatically. It should be possible to replace modules without switching off the entire machine to further minimize the changeover time.

### Engineering system
The separation of physics and logic as well as the separation of the functional software in a separate application module is a must for this variant. On the one hand, this is the only way to ensure that in the event of a changed hardware configuration (as in the example when replacing the gluing station), the associated software can also be activated. On the other hand, the development, testing, and commissioning of the module is thus decoupled from the rest of the project and the expected effects of complexity reduction are also achieved in the software. If this is not possible, it would be a knock-out criterion for this modularization variant. This applies in particular to the safety components (Section 5.3.3).

Even if integrable modules are a possibility for *successive modularization* in this and the following scenario, it should be urgently pointed out that the formation of *autonomous modules* should always be the first choice. After all, even with a successive approach, it offers the best opportunity to start with a truly future-proof modular design of the machine.

## 5.1.3 Decentralized hardware is possible to a limited extent

### Delimitation

With the automation technology used, decentralized inputs and outputs can be operated with the required reaction time, but the field bus system is not sufficiently real-time capable for decentralized drives.

In this case too, the formation of an integrable module is possible in principle but with the restriction that an additional serial coupling is required for the decentralized drives (Figure 5.3). This is possible, for example, with a drive system as described in [2]. The calculation of the drive functions takes place via a field bus, while the decentralized inverters are then coupled via a special drive bus. The decentralized modules then only have the task of reading out the position encoder and controlling the drive motor. In this configuration, the field bus does not necessarily have to

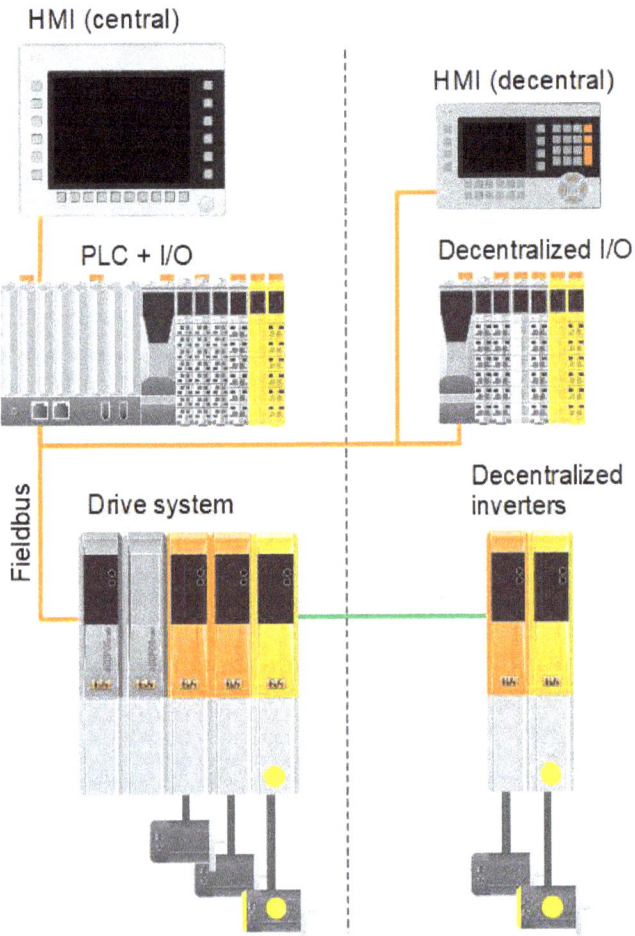

**Figure 5.3.** Automation with an integrable module and separate drive bus.

work highly dynamically, as it only has to transmit parameters and set values from the control system. The actual axis coordination takes place in the drive controller. Essentially, only the current setpoint and actual position values are transmitted via the additional drive bus.

The remaining structure for the connection of input and output modules and the HMI device as well as the boundary conditions specified there do not differ in principle from the scenario in Section 5.1.2, although the following aspects must also be taken into account:

**Field bus**

In the configuration shown, particular attention must be paid to the ability of a plug-in, which must now work for both connections.

**Engineering system**

The separation of physics and logic as well as the separation of the functional software in a separate application module also applies as a condition for this configuration. However, if the control project of the drive controller is not fully integrated into the rest of the project, two projects must be managed in addition to two bus systems. This can make the development, testing, and commissioning of the module more difficult or even represent a knock-out criterion for this variant.

## 5.1.4 Decentralized hardware is not possible

> **❗ Delimitation**
>
> The automation system cannot directly operate decentralized drives or distributed inputs and outputs with the required reaction time, but it does have interfaces for standardized communication.

If it is not possible to decentralize the input and output modules or the drive technology while guaranteeing the required dynamics, it is reasonable to rule out the possibility of forming an integrable module.

In practice, there are certainly different approaches to realizing the required dynamically demanding functions with additional hardware. For example, intelligent field devices or input/output modules with their own processor could be used, as discussed in Section 5.2.3.

Furthermore, the drive controllers could use an additional position encoder to detect the master position on the backbone itself and thus implement the decentralized functions with the required accuracy. However, it is questionable whether this additional effort is expedient, as it not only increases the complexity of the hardware and software but also incurs additional costs for the more expensive hardware. Furthermore, the mechanical installation of the position encoder alone results in additional integration costs.

Although an autonomous module can also initially generate higher costs for the required automation technology and design integration, this is offset by the aforementioned advantages of functional separation. Therefore, the only expedient solution under the given boundary conditions is the formation of an *autonomous module.*

Figure 5.4 shows a possible automation structure. The module now has its own controller and is linked to the controller of the rest of the machine via a standardized interface such as *OPC UA*, if possible in the real-time-capable variant *OPC UA TSN* (Section 5.4.3).

The following aspects must be taken into account in this configuration:

**Compatibility**
The input and output modules are again located directly on the backplane, but on the decentralized controller of the module, for which a type with less performance may now be sufficient. In the simplest case, this could be a functionally compatible CPU with less memory and/or a weaker processor but the same design. In this case, identical input and output modules could also be used. If the CPU has a different design and is not connection-compatible, other input and output modules may also be used, of which the functional compatibility must be checked.

In this case, it depends on whether the other technology is software-compatible if the control components from the same manufacturer are used and, if this is not the case, how much effort is required for the new software to be developed and how the engineering systems support export/import. In addition, the requirement to use as many identical parts as possible[5] can mean that the control components of the overall machine

---

[5]In order to negotiate better purchasing conditions and to minimize stock levels, spare parts lists, and administrative costs, manufacturers use as many identical components as possible for their different products. For example, two channels are left free in a 6-channel input module rather than using a slightly cheaper 4-channel module for a single case.

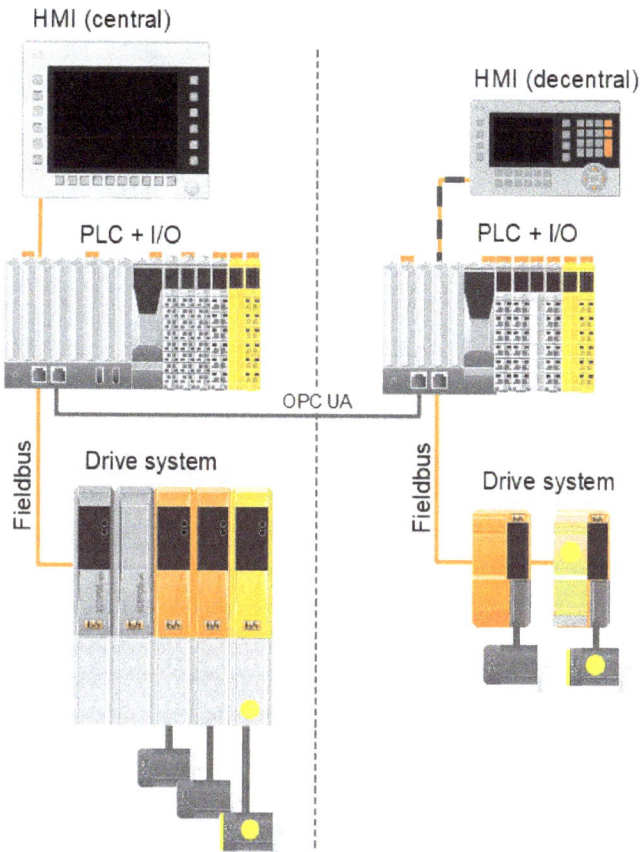

**Figure 5.4.** Automation structure with an autonomous module.

should also be used in the modular units, if possible, even if their perfor-mance is not required and they are more cost-intensive. This can even result in an economic advantage in the end if this reduces the initial outlay by not having to completely redevelop the relevant software components for the module and the price for the now larger quantities is reduced by additional discounts.

**Real-time capability**
Depending on the outsourced scope of functions, decentralization into an autonomous module will lead to a dynamic advantage due to fewer

control tasks to be processed or a less powerful CPU to be used, at least when using a type- and performance-identical controller. In order to assess this, suitable tests are recommended in addition to model calculations. Further options for ensuring high dynamic requirements in less powerful controllers are presented in Section 5.2.3.

The transmission of the synchronous master position, on the other hand, can be problematic. However, if the OPC UA TSN bus protocol is used, field bus cycle times of up to 50 $\mu s$ with a transmission jitter of < 40 ns are possible,[6] which is certainly sufficient for most processing machines.[7]

Additional wiring or additional mechanical tapping of the master position with a rotary encoder would only be conceivable if the dynamic requirements do not allow any of the above-mentioned leeway.

**Safety technology**
If a mechatronic unit is designed as an independently operational module, it must be subjected to a risk analysis in accordance with DIN EN ISO 13849 [1]. The same applies to the entire machine, as its safety concept is also changed by the removal or addition of a module. Each machine variant created in this way is an individual machine in terms of safety and must be individually tested, maintained and fully documented in the manufacturer's declaration. There are safety concepts that can also be used to certify modular machines and systems with an overall safety concept to make this easier. The topic is discussed in Section 5.3.1 and is particularly important in the scenario presented here.

**HMI components**
Let us first assume that the decentralized operating panel only has a browser as intelligence, is integrated into the module's own field bus, and communicates via OPC UA. There is also a web server on the decentralized controller, and the actual HMI project also runs there.[8] This constellation results in a very elegant concentration of the entire software on the module CPU in just one user memory.

---

[6] Corresponds to the current dynamic characteristic values as published in [24].
[7] More about OPC UA TSN in Section 5.4.3.
[8] Web-based HMI concepts are offered by almost all relevant manufacturers.

The result would be a simplified service and completely non-reactive development and integration. In addition, the central controller and main control panel can also use the services of the decentralized web server via the existing OPC UA interface, provided they work on the same principle. This means that with little additional effort, the operator has the choice of where to parameterize the adhesive dosing, for example.

When using an operating panel with its own intelligence and/or no web-based visualization, the entire communication and data management must sometimes be adapted and organized individually. Separate engineering systems may still be used for software development, with the control system being developed in one and the HMI project being developed in the other. This creates a new interface that has to be researched, designed, and documented thoroughly, which means additional work.

**Communication**
In addition to the aspects already mentioned, reference should also be made to the general design of the data interface of the decentralized system. A service-oriented interface creates the best conditions for a sustainable concept even with successive modularization, regardless of whether OPC UA with or without TSN specification. This applies both to communication within the machine and beyond. Remote maintenance, cross-system operation and visualization, machine safety, operation outside the machine (e.g. keeping the module ready in a docking station), as well as establishing I4.0 compatibility are all aspects that are much easier to manage with this form of modularization. All other protocols and procedures automatically lead to additional efforts and disadvantages in the next life cycle.

**Engineering system**
The porting effort is likely to be the lowest if there is a clean separation of physics and logic in the software project of the non-modular machine and the same automation system is also used in the new module. If the application can also be programmed in an object-oriented manner and the operating system of the controller is capable of multitasking, the conditions for effective and sustainable modularization are almost ideal. If this is not the case, the arguments in the following section apply.

### 5.1.5 Heterogeneous automation technology

> **!  Delimitation**
>
> Scenario 3 applies with the addition that the modular component is to be implemented by a supplier. The supplier uses its own automation system, which is incompatible with the system used in the rest of the machine but has standardized interfaces.[9]

This case deserves special attention, as the outsourcing and decentralization of subsystems is common practice in many companies. There are many reasons for this. There is either a lack of production capacity or there is no or insufficient know-how available for the technological sub-process within the company, for example, because the processing technology is new. However, economic factors can also simply play a role. Regardless of the reasons, there is no way around an autonomous module in this scenario.

First of all, the same technical boundary conditions must be observed as in scenario 3, except that they must be incorporated much more precisely into a specification sheet. After all, it makes a big difference whether the colleague across the table or a neighboring department is working on the sub-project or whether another company is taking on this job. In this case, agreements on delivery times, pricing, warranty, and service must also be made. This also means that the departments which might be involved in drafting the contract only know the title of the actual technological requirement.

However, these are precisely the aspects that can be mastered very well through function- and object-oriented modularization with autonomous modules. The appropriate instruments for this are a clear functional separation of the module and the rest of the machine or system, a service-oriented communication via standardized interfaces, a web-based HMI technology, and the use of integrated safety technology.

### 5.1.6 Summary

A step-by-step approach is certainly practicable even though a comprehensive and thorough function- and object-oriented modularization should be

---

[9]This scenario also applies if technologically required components such as laser sources, hydraulic units or other supplier components with their own automation technology are to be integrated into machines.

the first choice. If the technical, economic, and organizational framework conditions allow it, a solution with autonomous modules offers the greatest potential in terms of complexity reduction and sustainability.

## 5.2 Real-time capability decentralized systems

In numerous industries, there are more or less time-critical processes whose precise processing is crucial for product quality and production speed. The following applications are examples of this.

### Application: Print mark recognition

In a *packaging machine*, fast-moving markings on the packaging material — e.g. print and product identification marks — must be recognized and measured. The subsequent packaging process is synchronized to the positions determined. If this is to be faster and more precise, measurement is typically required with a resolution of a few microseconds. As a result, productivity is increased, and incorrect packaging is avoided.

### Application: Register control

The quality of multicolored print products crucially depends on how synchronously the different printing units in a *printing press* work in relation to each other. Register marks are printed on the edge of the page in each color and then measured in order to control this process. The measurement result is then a parameter for the position control of the subsequent printing cylinders. High-precision measurement and control in the single-digit $\mu s$ range reduces the consumption of printing inks and paper by avoiding waste.[10] Until now, this has required expensive special hardware.

### Application: Plastic injection molding

Extremely fast signal processing and maximum accuracy are essential for controlling and regulating the injection process in *plastic injection molding machines*. The precision of the switchover from speed to pressure control has a direct influence on product quality and the efficiency of raw material utilization (minimum drying time).

Regardless of the application, only exact compliance with the reaction time required for the technological process guarantees the desired product quality in the end. It does not matter whether these times are in the range

---

[10]This is the term used in the printing industry for rejects.

of seconds or microseconds, the requirements for compliance with them in a technological process are always the same.

The specified tolerances exist in every process. It also does not matter which automation technology and which industry is being considered. If, for example, field devices with wireless data communication are to be used in a refinery for the distillation of crude oil in a reactor cascade, then the automation engineer has to deal with the same requirements as his colleague who works on register control in a printing machine, although the required reaction times of both applications are several orders of magnitude apart.

This section shows which influencing factors have an effect and how, and which concepts can be used to influence the reaction time and how. The focus here is not on the calculation, modeling, or tuning of control loops and control processes.

There are numerous standard works on this subject that belong on the bookshelf of every automation engineer. Rather, the focus is on those considerations that have an influence on the direct design of the automation system in order to achieve the best possible values for the quality-oriented fulfillment of the technological functions. These include not only the time dimensions but also aspects such as stability and reliability as well as the reduction of development and commissioning costs.

In modular mechanical engineering in particular, the reaction time of an automation system is slowed down by the dynamic resources of the controller, field bus, and infrastructure. In the following, we therefore want to develop a basic understanding of the individual factors that have a greater or lesser influence on the reaction time in modular structures. Finally, we show possible solutions for reliably meeting high dynamic requirements.

### 5.2.1  Reaction time: Definition and requirements

In simplified terms, the reaction time is the time a system needs to react to a process event. If a change of state is detected at a signal input, a certain amount of time will pass before a change of state calculated in the program occurs at a signal output. However, this reaction time depends on many factors and is determined by the technology. Typical process requirements are, for example, the following guide values:

- $T_{\text{Reaction}} > 1\ s$
  - ○ Chemical processes, bioreactors
  - ○ Temperature control in industrial ovens

- $100\ ms > T_{\text{Reaction}} > 10\ ms$
  - ○ Speed control in conveyor and feeding technology
  - ○ Temperature zone control in extruders

- $10\ ms > T_{\text{Reaction}} > 1\ ms$
  - ○ Standard PLC functions

- $T_{\text{Reaction}} \ll 1\ ms$
  - ○ Position control for servo drives
  - ○ Print mark control in packaging machines
  - ○ Edge detection for print and post-print processes
  - ○ Register control in printing machines
  - ○ Control of the injection pressure in plastic injection molding machines
  - ○ Intelligent cam switch units

Figures 5.5 and 5.6 show in a highly simplified representation which factors determine the reaction time in a centralized and decentralized control structure.

If the sensor of a field device (light barrier, initiator, analog measured value, etc.) registers a change of state, this information must pass through the electronics of the field device and the input module to the control

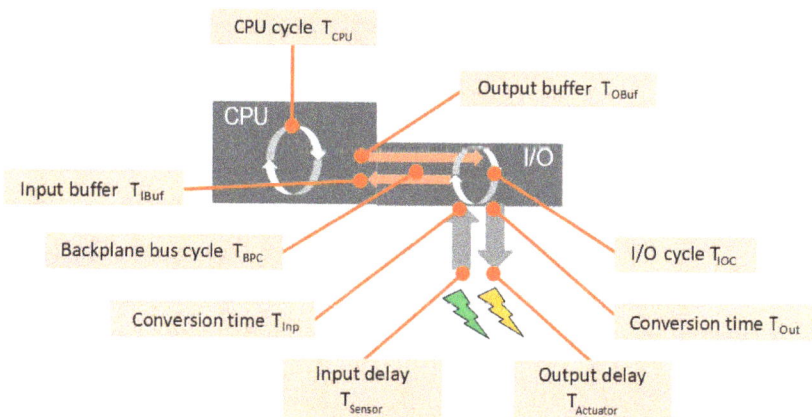

**Figure 5.5.** Elements of the reaction time in a central structure.

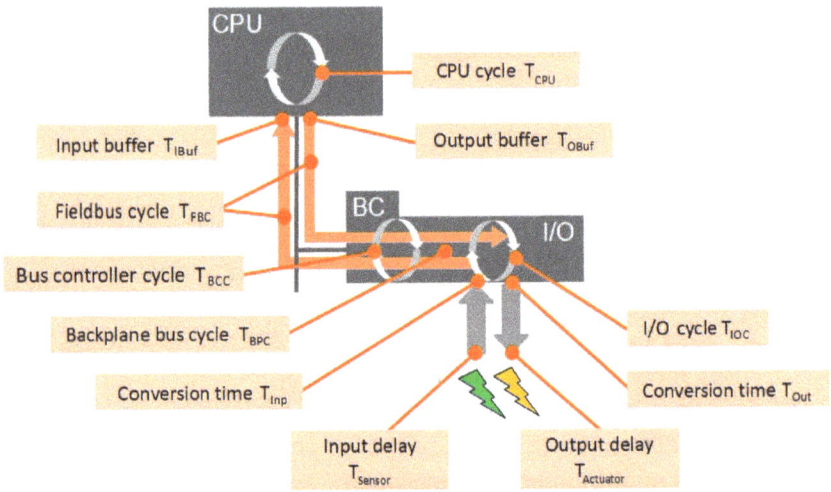

**Figure 5.6.**   Elements of the reaction time in a decentralized structure.

CPU. If an output action is calculated during the processing of this information, the corresponding command must be sent to the electronics of the output module and on to the actuator. Along the way, the information passes through several interfaces that differ both in number and in their behavior, depending on the structure of the automation technology.

From this signal curve, the reaction time in a central structure according to Figure 5.5 is in principle the simple sum of all time components in the order in which they occur:

$$T_{\text{Reaction}} = T_{\text{Sensor}} + T_{\text{Inp}} + T_{\text{IOC}} + T_{\text{BPC}} + T_{\text{CPU}} + T_{\text{BPC}} + T_{\text{IOC}} + T_{\text{Out}} + T_{\text{Actuator}} \quad (5.1)$$

If the input and output modules operate with the same cycle time, the following applies:

$$T_{\text{Reaction}} = T_{\text{Sensor}} + T_{\text{Inp}} + 2 \cdot T_{\text{IOC}} + 2 \cdot T_{\text{BPC}} + T_{\text{CPU}} + T_{\text{Out}} + T_{\text{Actuator}} \quad (5.2)$$

With this assumption, the decentralized structure according to Figure 5.6 applies in principle:

$$T_{\text{Reaction}} = T_{\text{Sensor}} + T_{\text{Inp}} + 2 \cdot T_{\text{IOC}} + 2 \cdot T_{\text{BPC}} + 2 \cdot T_{\text{FBC}} + T_{\text{EPuf}} + T_{\text{CPU}} + \\ T_{\text{APuf}} + T_{\text{Out}} + T_{\text{Actuator}} \quad (5.3)$$

The meaning of the individual time components is summarized in Table 5.1.

**Table 5.1.**   Elements of the reaction time at a glance.

| Description | | Meaning | Typical values |
|---|---|---|---|
| Input delay | $T_{Sensor}$ | – Signal propagation time from acquisition at the transducer to output at the field device terminal<br>– Also includes the signal processing of intelligent field devices | Depending on the device |
| Output delay | $T_{Actuator}$ | – Runtime of the signal from provision at the terminals of the field device to the action<br>– Also includes the signal processing of intelligent field devices | Depending on the device |
| Conversion time | $T_{Inp}, T_{Out}$ | – Analogue-to-digital or digital-to-analogue conversion of the sensor data in the input or output module | Depending on resolution and data type |
| Cycle time I/O-Modul | $T_{IOC}$ | – Data pre-processing such as filtering, smoothing, Fourier analysis, etc.<br>– Transfer of the pre-processed values to the backplane bus<br>– Transfer of the output data from the backplane bus<br>– Transfer of any further processed values to the signal converters | Depending on parameters used |
| Cycle time backplane bus | $T_{BPC}$ | – Serial transmission from the I/O modules to the bus controller | $\leq T_{FBC}$ |
| Cycle time bus controller | $T_{BCC}$ | – Organization of data traffic between field bus and backplane bus | $= T_{BPC}$ oder $T_{FBC}$ |
| Cycle time field bus | $T_{FBC}$ | – Data transmission between the bus devices in real time | $>200\ \mu s$ |
| Input and output buffer | $T_{IBuf}, T_{OBuf}$ | – Receiving or providing data from or to peripherals | Depending on device and field bus type |
| CPU-Cycle time | $T_{CPU}$ | – Cyclical processing of the user applications | $>500\ \mu s$ |

A look at the column of typical values reveals that many of these elements cannot be represented in a typical value range. For example, the *input delay* of a sensor is explicitly dependent on the type of measuring element, the electronics of the field device, and its parameterization. The same also applies to other time components, so the first priority must be the careful selection of the appropriate device technology. The *conversion time*, on the other hand, often depends on the technologically required parameterization, just like filter constants and the like, and must be looked up in the documentation. However, there are also elements that cannot be found in any manual or online forum because they are part of the manufacturer's know-how. These include the cycle times of the backplane bus as well as the provision times of the data in the input and output buffer of a CPU or a bus controller.

For this reason, we only want to focus on the conversion time components that are particularly important when switching from a centralized to a decentralized structure or that can be influenced by parameterization and selection. Assumptions are made as follows[11]:

### Input and output buffer of the CPU ($T_{IBuf}$, $T_{OBuf}$)

The way in which data is provided in the CPU of a controller varies greatly from system to system and also depends on whether the data is exchanged via a backplane bus or field bus. In most systems, the backplane bus is handled by the operating system in the background and parallel to processing the application. If it is a field bus, however, the behavior varies depending on the integration. If the interface is integrated directly on the CPU, a driver of the operating system will take care of receiving, decoding, and distributing the data contained in the telegram. This is also usually done in the background and does not require any additional time. If, however, the field bus is integrated via a separate interface module, the bus controller must process the data packet and then send it to the CPU via the backplane bus. At least one additional cycle time of the bus controller must be taken into account in this constellation, depending on the manufacturer's design. Considering these values, which cannot be

---

[11] The values mentioned are the best-kept know-how of the manufacturers and are generally not published. The following assumptions reflect the author's own experience as well as verbal statements made by some manufacturers at trade fairs, conferences, and similar events.

determined exactly, the time components mentioned should be neglected or integrated into the field bus cycle.

> **Assumption 1**
>
> The time components $T_{IBuf}$ and $T_{OBuf}$ are assigned to the field bus cycle $T_{FBC}$.

## Cycle time I/O module ($T_{IOC}$)

In principle, input/output modules only have the task of receiving input signals or transmitting output signals to the actuators. In addition to a cyclically operating processor,[12] they have the appropriate hardware to process or generate the mostly standardized signals or to operate a special serial interface such as *I/O-Link*. These modules have various filters that can be parameterized as required in order to eliminate signal transmission interference such as noise or contact bounce in an input module. If a filter constant is 25 ms, for example, a signal change is only detected if the pulse lasts longer than 25 ms. If, on the other hand, a software filter is not used, the input signal can also only be filtered by the hardware, and this can be just a few microseconds or even nanoseconds for high-speed inputs.

All time components within an I/O module should be combined to $T_{Inp}$ or $T_{Out}$ because different amounts of time are also required for the conversion from or to analog signals depending on the data format and required accuracy (from a few microseconds for high-speed to a few tenths of a second for temperature inputs) and can only be influenced by appropriate selection or parameterization.

> **Assumption 2**
>
> The cycle time of the I/O module $T_{IOC}$ is assigned the time components $T_{Inp}$ and $T_{Out}$, which are determined by the hardware used and can also be parameterized individually in some cases.

---

[12] In modern systems, so-called *field programmable gate arrays* (FPGA) are usually used for this purpose.

### Cycle time backplane bus ($T_{BPC}$)

Data transfer via the backplane bus of a controller or a bus node is usually highly efficient and extremely fast. In addition, the transfer is usually synchronized with the cycle of the CPU and bus controller so that there is only an insignificant time delay at this point.

> **! Assumption 3**
>
> In a centralized structure, $T_{BPC}$ is neglected and considered a decentralized component of $T_{FBC}$.

### Cycle time bus controller ($T_{BCC}$)

Bus controllers generally work synchronously with the field bus cycle and communicate with the assigned I/O modules in parallel via the local backplane bus. It depends on the system and should be enquired about if necessary whether and to what extent this results in additional time delays.

> **! Assumption 4**
>
> The time component $T_{BCC}$ is added to the field bus cycle $T_{FBC}$.

### Cycle time of the CPU ($T_{CPU}$)

At first glance, it is quite simple how a CPU actually processes the user programs. Inputs and data of the memory are read and processed, the results are written to the memory or sent to the outputs, and after a few routines of the firmware, a new start is made. The resulting cycle requires the time $T_{CPU}$. It depends on the length of the program to be processed and the load from the operating system and how long this time is. However, the example of a *skew control* used in Section 5.2.3 shows us that there are functions in almost every application that have different dynamic properties and therefore need to be processed at different speeds. Nevertheless, simple linear program processing from the first to the last command results in the same cycle time for all functions. The individual manufacturers solve this quite differently with timers, interrupts, or multitasking. If, for example, a faster application is required, a timer can be parameterized in some systems, which interrupts the main program at a fixed cycle and forces immediate processing. The disadvantage is obvious, as constant interruptions make the reaction time for the other processes longer and longer or unpredictable. This is particularly the case if the interruptions do not occur in a fixed cycle but are triggered by external signals. This reduces the control quality of the other processes, and an extensive project with many time-critical interruptions makes the system unstable and prone to errors.

A better alternative is the use of *multitasking systems*, in which various dynamic processes are assigned to corresponding tasks or task classes. The advantage here is that the cycle times of the individual tasks do not change if the parameters are set correctly, and a stable system is created. We want to assume this method of working for the following considerations:

### ! Assumption 5

The cycle time $T_{CPU}$ stands for the processing time of an application program in the CPU and is constant.

### Cycle time of the field bus ($T_{FBC}$)

First of all, it should be assumed that the data traffic on the field bus is asynchronous to the coupled components. This means that the cycles of the controller, bus controllers, and field devices are not synchronized with each other. This assumption has practical significance insofar as current Ethernet-based bus systems such as *Modbus TCP* or *Profinet RT* do not have a synchronization mechanism like *POWERLINK* or *EtherCAT*. The transmission times of these bus protocols are therefore subject to the conditions of Ethernet physics, in which collisions and the behavior of switches stochastically increase the time required for data transport and thus generate additional jitter.

### ! Assumption 6

The cycle time of the $T_{FBC}$ field bus is not constant, and transmission is asynchronous.

The time diagrams resulting from these assumptions are shown in simplified form in Figure 5.7 for the centralized structure and in Figure 5.8 for the decentralized structure.

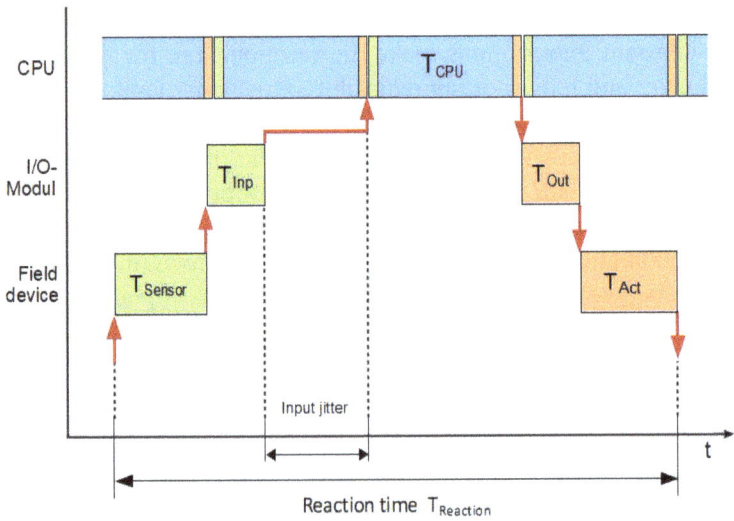

**Figure 5.7.**   Principal signal path in a central structure.

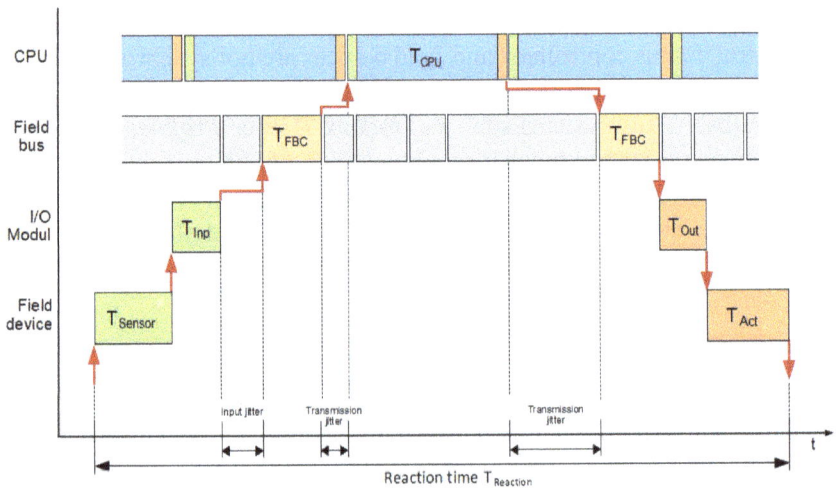

**Figure 5.8.**   Basic signal path in a decentralized structure and asynchronous data.

The signal curve in Figure 5.8 clearly shows the undesirable effects of asynchronous data transmission. In this constellation, it is almost impossible to predict how long the cycle time will actually be, as the completely asynchronous operation of the field bus results in incalculable time delays at the transfer points. If necessary, this can only be dealt with by iterative parameterization, intensive testing, and adjustments to the network infrastructure. Synchronized operation of servo axes is either not possible at all with these properties or only possible with a loss of dynamics.

The biggest problem here is the jitter that occurs, which we address in the following section. We then present possible solutions for ensuring a significantly more stable and, above all, shorter reaction time.

### 5.2.2  Jitter: The great unknown

In translation, jitter means flickering, fluctuation, or instability. In technology, jitter is manifested by clock jitter and uneven signal transmission over time.

This effect is particularly evident with *Voice over IP* (VoIP) or streaming video files, for example. In these applications, it is important that the data packets are played back very synchronously. If this is not the case, we notice this through choppy audio or video.

To avoid this *transmission jitter*, a buffer is built up in the receiver before playback starts. The clip is not played until this buffer is sufficiently full, and while we watch the video without disturbing jitter, the receive buffer is continuously refilled in the background. This allows short delays in data transmission, for example, due to slow or overloaded Internet, to be bridged more or less well. The playback only stops when the interruptions become so long that the buffer content has been played back.

Unfortunately, this method does not work in automation technology. After all, from a control system, we expect that the elaborately calculated controller can work reliably with the required actual values and that the electronically coupled servo drives move absolutely synchronously with the master drive. Various methods have been established to ensure this, which are discussed later on. But first, let's take a look at the causes of jitter in control technology.

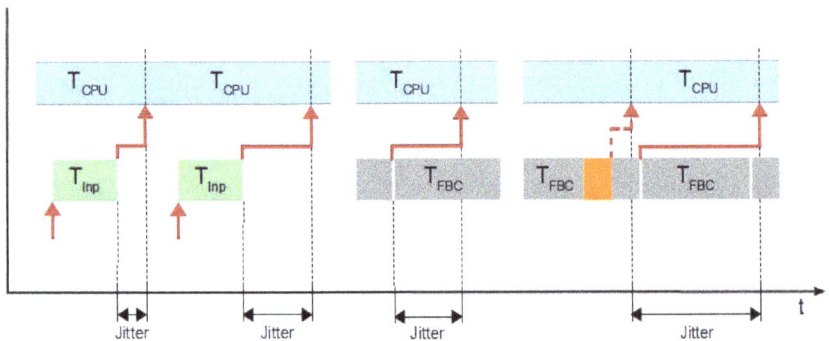

**Figure 5.9.**    Typical phenomena of the occurrence of jitter in control technology.

Figure 5.9 shows three typical scenarios. Let us first look at the signal transmission from an input module to the controller on the left-hand side of the picture.

It is noticeable that the state change of an input signal practically never occurs exactly at the start of a CPU cycle. Rather, a signal arrives in the controller at some point during a cycle and has to wait until a new one begins. Whether the information spends this waiting time in the input buffer of the CPU or is held on call by the input module is irrelevant. It is in the nature of things that there is no practicable solution for synchronization with a sporadically occurring signal. This leaves the challenge of incorporating the unavoidable *input jitter* into the calculation of the reaction time, even if it can be unbearably long for automation technology with a maximum length of one CPU cycle.

The illustration in the middle of Figure 5.9 shows the occurrence of a jitter, which is caused by a field bus operating asynchronously to the CPU clock. The data packet of the bus protocol ends up in the input buffer of the controller at some point during a CPU cycle and must wait there for a maximum of once the time $T_{CPU}$ until it is processed. As shown in Figure 5.8, jitter can also occur during data transfer from the CPU to the field bus, and Figure 5.9 on the right shows how the jitter can change significantly from one cycle to the next due to transmission errors or interruptions.

Therefore, when using asynchronously operating field buses, it is not possible to use the *transmission jitter* for each transition, at least one cycle time $T_{CPU}$ or $T_{FBC}$ must be taken into account in the calculation of the reaction time. However, $T_{FBC}$ in Figure 5.8 may be difficult to determine so that in this case only rough estimates or results from long-term measurements can be used.

Taking into account Assumptions 1–6 and the inclusion of input and transmission jitter, the following are the results for the calculation of the reaction time in a central structure according to Figure 5.7:

$$T_{\text{Reaction}} \leq T_{\text{Sensor}} + T_{\text{Inp}} + 2 \cdot T_{\text{CPU}} + T_{\text{Out}} + T_{\text{Actuator}} \qquad (5.4)$$

The following applies to the decentralized structure according to Figure 5.8:

$$T_{\text{Reaction}} \leq T_{\text{Sensor}} + T_{\text{Inp}} + 4 \cdot T_{\text{FBC}} + 2 \cdot T_{\text{CPU}} + T_{\text{Out}} + T_{\text{Actuator}} \qquad (5.5)$$

The jitter to be taken into account in the central structure is therefore

$$T_{\text{Jitter}} \leq T_{\text{CPU}} \qquad (5.6)$$

and in a decentralized structure

$$T_{\text{Jitter}} \leq T_{\text{CPU}} + 2 \cdot T_{\text{FBC}} \qquad (5.7)$$

not yet taking into account the fluctuations of $T_{\text{FBC}}$.

In the following, we show which measures can be used to initially stabilize the reaction time by avoiding or reducing the jitter.

**Input jitter**
In the left section, Figure 5.10 shows that input jitter cannot be eliminated, but its time dimension can be reduced by more frequent sampling and the synchronization of the CPU and field bus cycle. This solution is practical if the CPU performance is sufficient for the other processes due to the higher load. The interrupt-based solution shown on the right can indeed eliminate the input jitter, but jitter is generated or amplified at other points (Section 5.2.3).

**Transmission jitter**
As the interrupt variant on the right in Figure 5.10 shows, the method can also be used to eliminate transmission jitter.[13] It depends solely on the

---

[13] In practice, a few microseconds remain because field bus and CPU must recognize the request and interrupt running processes.

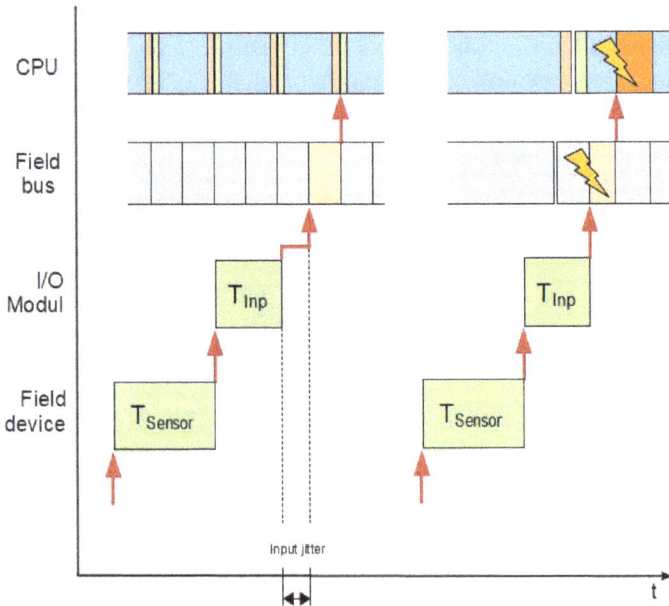

**Figure 5.10.**   Reduction of the input jitter through shorter, synchronized sampling time or interrupt.

technological requirements whether this principle or a fixed field bus cycle is the better option. In a processing machine, other factors play a role than in a building automation system with a widely branched network and many decentralized participants. When a storm is approaching, it is completely irrelevant if the command to retract all shutters is executed one second earlier or later. In a machine with electronically coupled, highly dynamic servo drives, on the other hand, stochastic transmission delays of just a few microseconds can have a negative impact on the control quality.

There are also methods that tolerate jitter by providing the user data with time stamps before transmission and then interpolating the actual values to the time of use in the receiver. However, this method can fail, especially in highly dynamic applications, because the speed changes to the same extent during transmission in the event of strong acceleration, for example, and extrapolated position values inevitably deviate. The result is a deterioration in the running smoothness of the drives involved and thus increased mechanical wear.

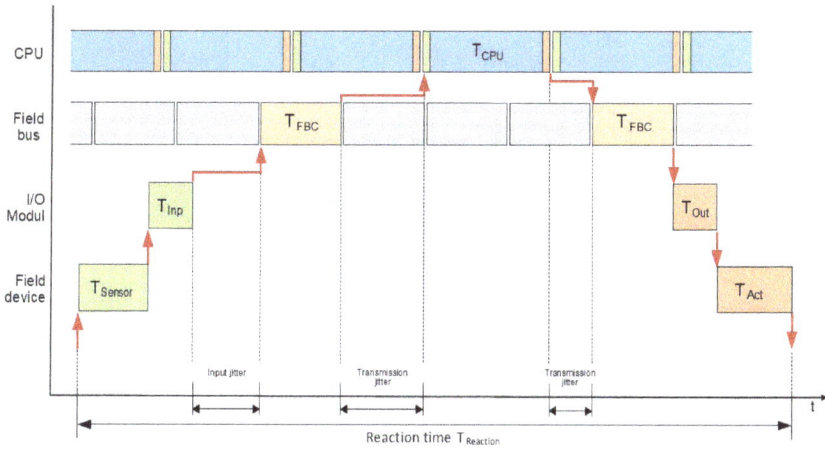

**Figure 5.11.** Principal signal curve in a decentralized structure and asynchronous data transmission with a fixed field bus cycle.

Of all solution approaches, the elimination of all avoidable interference points where jitter can occur is certainly the best choice for largely jitter-free data transmission. In the first step, this can be a constant field bus cycle, which does not have much effect in terms of dimension (in principle, relationships 5.5 and 5.7 still apply) but which at least makes the reaction time more predictable overall (Figure 5.11).

A significant improvement only occurs if the cycles of the CPU and field bus are synchronized. This possibility is offered, for example, by the field bus *POWERLINK* [3] with the high-precision synchronization pulse *Start of Cycle* (SoC). All bus participants can synchronize to this signal and thus allow the transmission to run practically jitter-free (Figure 5.12).

As a result, the total jitter is only at the level of the input jitter at

$$T_{\text{Jitter}} \leq T_{\text{CPU}} \tag{5.8}$$

and the reaction time

$$T_{\text{Reaction}} \leq T_{\text{Sensor}} + T_{\text{Inp}} + 3 \cdot T_{\text{FBC}} + T_{\text{CPU}} + T_{\text{Out}} + T_{\text{Actuator}} \tag{5.9}$$

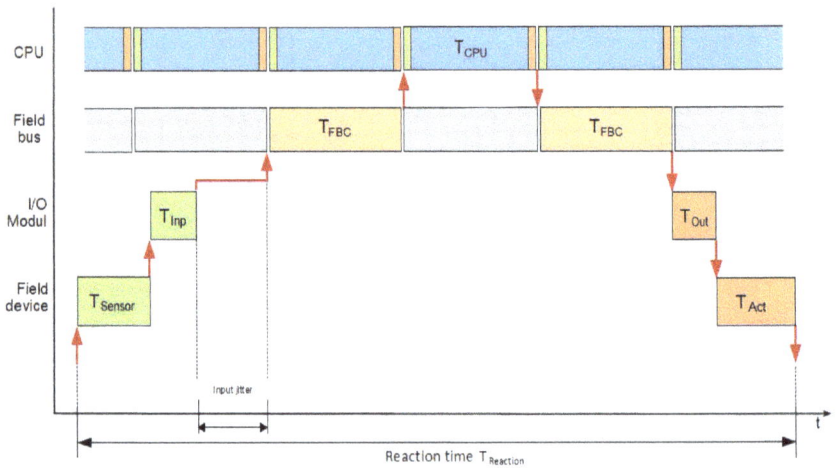

**Figure 5.12.** Principal signal curve in a decentralized structure with synchronization of field bus and CPU cycle.

With $T_{FBC} = T_{CPU}$, the result is

$$T_{Reaction} \leq T_{Sensor} + T_{Inp} + 4 \cdot T_{CPU} + T_{Out} + T_{Actuator} \tag{5.10}$$

Although the illustration shows how the transmission jitter can be eliminated, it also makes it clear that the reaction time in a decentralized structure can increase by up to two CPU cycles compared to a centralized structure due to the required data transmission. It is precisely this fact that can lead to problems in the control quality and thus the quality of the execution of technological functions when technological functions are separated from a central control system into a modular system, regardless of whether it is integrated or autonomous.

In order to minimize or completely avoid this influence, there are solutions that we now look at.

### 5.2.3  Short response times in decentralized structures

To demonstrate this problem and the different approaches to solving it, let us take a simple example from brochure production. Figure 5.13 shows the functional principle of skew detection as used in many, sometimes

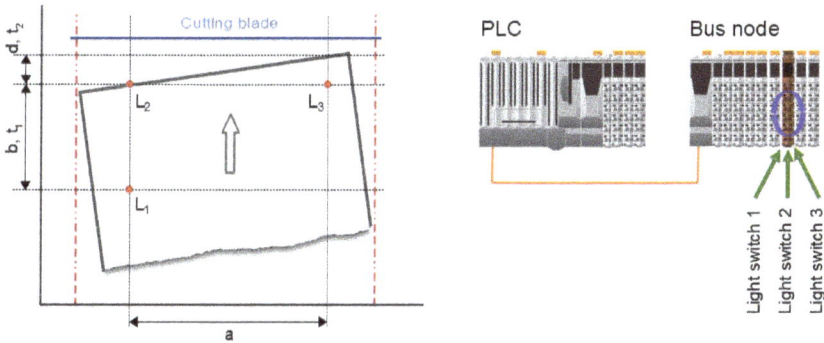

**Figure 5.13.**   Functional principle and automation diagram for skew detection.

very simple machines. Examples of this are the cut control of the three-side trim of a perfect bound brochure or the cutting of cardboard for the production of book covers.

The basic functional sketch shows how the product to be measured can twist in the lateral guides if it is adjusted incorrectly. This can sometimes only be a few tenths of a degree, but depending on the quality specification, this would lead to a rejection of the product. The operator must therefore monitor the frequency of excessive skewing on the machine and, if possible, the extent of the deviations in order to be able to make the appropriate fine adjustments.

In the simplest case, an skew position can be determined with the design shown in the picture by means of an antiviolence check of the light scanners $L_2$ and $L_3$. If both are covered at the same time, there is no skew, otherwise the product is ejected. If the degree of deviation $d$ is also to be measured, two approaches are possible.

If a synchronous position signal is available, e.g. from the transport system, both light scanners can trigger the respective current position. The difference between the two values can be used to easily calculate the degree of inclination $d$ using constant $a$. If, on the other hand, no position is available or is not available with the required accuracy, the product speed can be determined directly using the constant $b$ and the time difference $t_1$, which covers the light scanners $L_1$ and $L_2$.[14] If it is now assumed that the speed is uniform, which is very likely to be the case, then a qualitative statement can also be made using this value, the constant $a$, and the time difference $t_2$.

---

[14] This would be a typical solution for an autonomous module.

What looks very simple in theory poses a few hurdles in practice. Even with the simple antiviolence test, the question of how short or, better, how long the signals must be in order to be recognized in the control system must be clarified. If we look at the situation in a saddle stitcher for the production of wire-stitched brochures, a product moves through the measuring device at a speed of 2.2 m/s at 14,000 copies per hour,[15] and the cutting angle can only deviate by a maximum of ±0.5°.[16] For a = 100 mm, this means a $t_{2,\text{good}}$ of ≈ 400 $\mu$s. Assuming that the CPU and field bus operate synchronously and with the cycle time $T_{\text{CPU}} = T_{\text{FBC}} = 1$ ms — which is quite sporty in practice — in a central structure, the sensor signals can also only be detected and evaluated at 1 ms intervals regardless of the jitter. Figure 5.14 clearly shows the resulting consequences.

Depending on the distance from the scanning time at which the signal edges occur, misinterpretations can occur. A good product position is detected as bad in the left part and a faulty one as good in the right part. The problem has been known for as long as there have been digital systems, and it can be solved quite simply according to the *Nyquist–Shannon sampling theorem* by a higher sampling rate, with at least twice the frequency. With a CPU cycle time of $T_{\text{CPU}} \leq \frac{1}{2} \cdot t_{2,\text{good}}$, the *good* and *bad* states can be recognized much better.

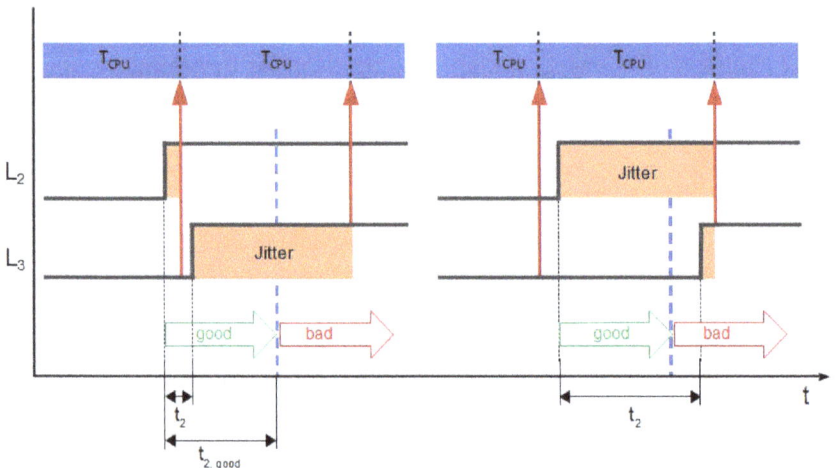

**Figure 5.14.**   Possible misinterpretations due to excessive scanning time.

---

[15] Typical values for a medium capacity saddle stitcher.

[16] With an A4 format, the resulting deviation of 0.5 mm is almost waste.

The example shows us that the detection of very short pulses is already a challenge for a centralized structure; in a decentralized arrangement with an involved field bus, this is even less possible. It also becomes clear that even in seemingly simple processes, highly dynamic signals have to be processed, and the dynamic requirements for processing just one signal can make thorough modularization much more difficult or even fail.

Now that we have recognized the problem, we want to discuss some possible solutions and how they can be implemented in modular but also central concepts.

### 5.2.3.1 *Increase in system performance*

The observations in Figure 5.14 have shown that signal states must be scanned at least twice as fast for detection and processing with the required technological accuracy in cyclically operating systems. This means that both the control hardware and software as well as the field bus must be designed accordingly. Figure 5.15 shows the effect of more powerful data transmission alone.

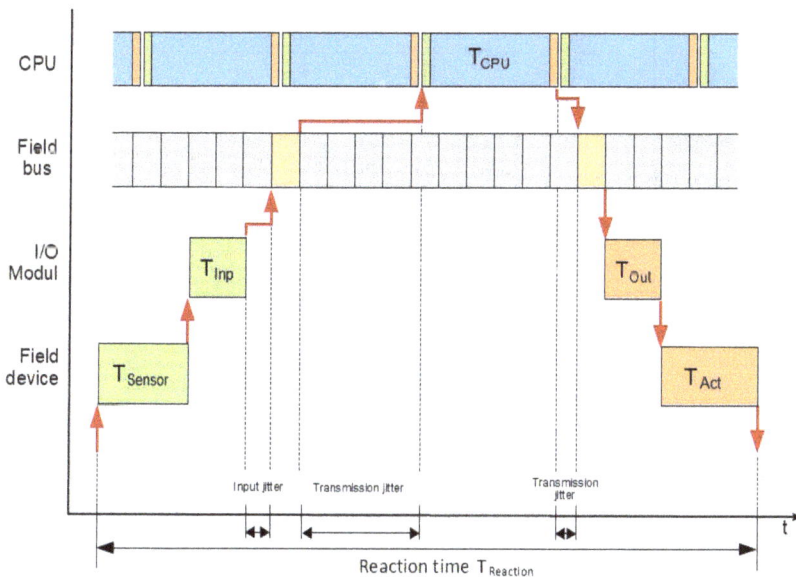

**Figure 5.15.**   Shortening the reaction time through high-performance, asynchronous data transmission.

In principle, the *Nyquist–Shannon sampling theorem* also applies in this case, in that the signal is forwarded much more frequently in relation to a CPU cycle.

If the overall system performance is increased, $T_{\text{Reaction}}$ can be shortened even further because in this case the dominant time component $T_{\text{CPU}}$ is also reduced and if the cycles of the CPU and field bus are also synchronized, there is no jitter (Figure 5.16). This means that $T_{\text{CPU}}$ alone dominates the reaction time, while the time component of the data transmission loses influence, and the dynamic conditions of a central structure are almost achieved. The stability of the reaction time achieved at the same time would be a further advantage of this concept.

Even if the solution approach seems quite simple and plausible, it is not a panacea in practice. In order to realize the CPU cycle time of $T_{\text{CPU}} < 200\ \mu s$ required in our application case of *skew detection*, a high-performance CPU would be required just for this function, which would go beyond almost any budget in this dimension.[17] This does not even

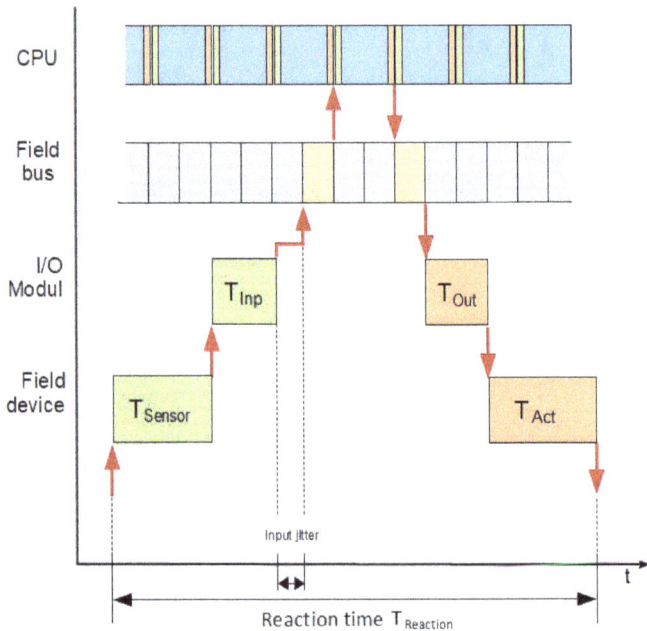

**Figure 5.16.** Principal signal curve in a decentralized structure with synchronous transmission at high system performance.

---

[17]If one also assumes that the control hardware in an autonomous module must be less powerful due to a lower workload, then this approach seems downright absurd.

consider how the quality limit can be parameterized. With such an approach, this can either only take place within a narrow framework, or the CPU cycle time must be adjusted for higher accuracy requirements. However, this raises the question of what impact this has on all other processes. The situation becomes even more obvious if the inclination is to be measured quantitatively and must be sampled much more finely, which leads us to the following conclusion:

> ☑ **Increase in system performance**
>
> – Response times well below 1 ms are not realistic in decentralized systems, even with high system performance
> – The basic system behavior is not affected
> – The use of very powerful hardware is cost-intensive
> – The data transmission infrastructure requires greater effort
> – The I/O scope cannot be expanded at will
> + The software can be developed flexibly and with standard tools

### 5.2.3.2 *Interrupt based systems*

The use of interrupts has already been considered as a possible measure to reduce the input jitter. Figure 5.17 shows how this method can be used to shorten the reaction time overall and which side effects must be taken into account.

First of all, it must be noted that the goal of drastically reducing reaction time can only be achieved with this concept for the functions that are also designed using this method. This can also work in the example of skew detection if the light scanners each trigger an interrupt, the information is transmitted immediately and is also processed in the CPU in an interrupt service routine (ISR) without delay. Depending on the hardware equipment, reaction time $s$ of a few microseconds can be achieved.

This approach is particularly interesting in small systems with many non-time-critical tasks and only a few highly dynamic functions. This approach is therefore also ahead in economic terms.

However, it also sets technological limits, as can be seen in Figure 5.17. Each interrupt generates additional and, above all, very unpredictable jitter for all other functions.

Furthermore, synchronization is not possible in interrupt-based systems, as this would be constantly disturbed. As a result, there is input jitter *plus* transmission jitter *plus* jitter caused by interrupts. If you now imagine

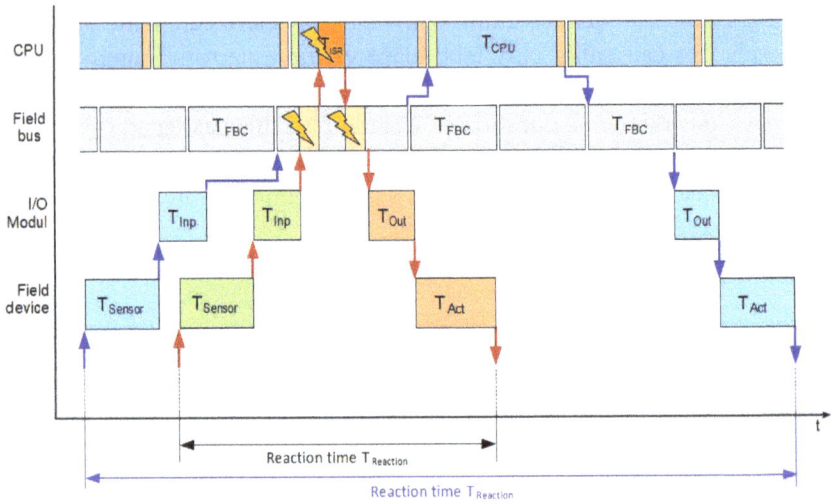

**Figure 5.17.** Principal signal path in a decentralized structure with interrupt-based transmission and processing.

that many functions (have to) work with this method, the result is an extremely unstable system with significantly increased testing and commissioning costs.

## ✓ Increase in system performance

- Due to a lack of synchronization, data transmission is asynchronous and heavily jittered
- Interrupts cause additional incalculable jitter for the rest of the system
- The scope of interrupt-based functions is limited and cannot be extended at will
- The system tends to be unstable, especially with numerous interrupt-triggering functions
- Increased testing and commissioning effort
+ The software can be developed flexibly and with standard tools
+ Depending on system performance, response times of a few microseconds can be achieved

### 5.2.3.3 *Intelligent field devices*

Intelligent field devices make it possible to implement a wide range of functions outside of the control system. To this end, manufacturers have equipped their field devices with powerful processors and provided them with software functions that can, for example, independently perform the functionality of *skew detection* by means of precise and highly accurate edge detection [4]. For this purpose, they have optimized inputs and/or outputs and also behave like an independent autonomous module. Via the field bus or other serial interfaces,[18] they make the processed data available to the controller, where it is processed less time-critically within the rest of the program sequence. The diagram in Figure 5.18 shows the basic structure of this concept.

The field device works completely autonomously in its own cycle after the desired functionality is selected and parameterized asynchronously from an available function pool during the initialization phase. During operation, the results are transferred to the controller as status information with correspondingly lower dynamic requirements. Figure 5.19 shows the

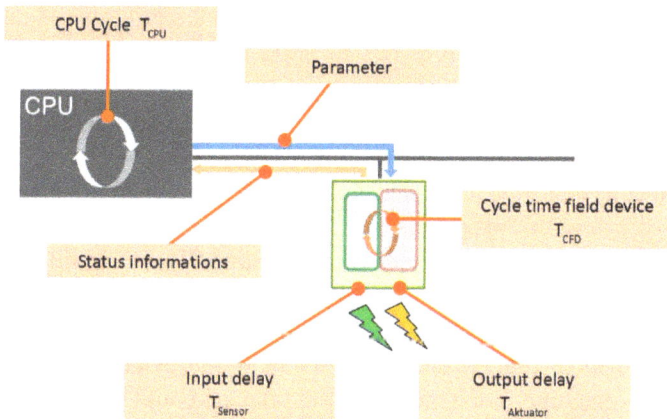

**Figure 5.18.** Elements of response time with intelligent field devices.

[18] IO-Link can also be used to connect such field devices, where additional information and parameterization can be transmitted serially in addition to the sensor signals [42].

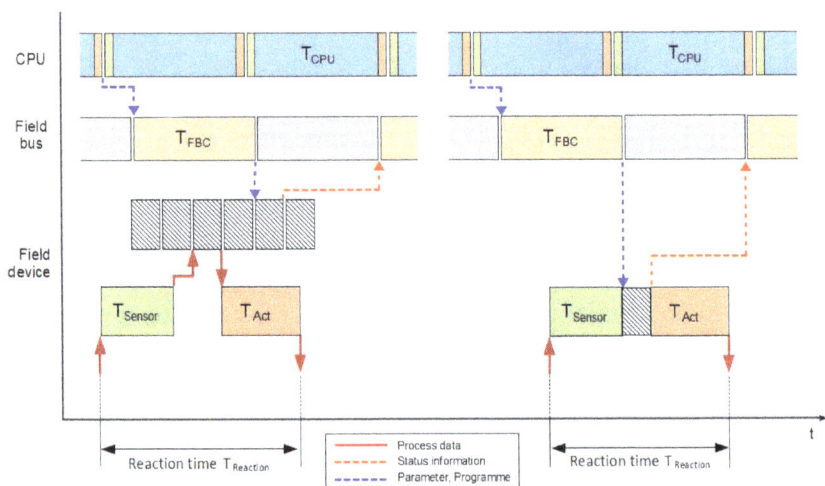

**Figure 5.19.**   Principal signal curve in a decentralized structure with intelligent field devices.

time sequence for synchronous data transmission in two different operating modes. The field device on the left works in a similar way to a fixed-cycle controller and therefore has to cope with the effect of the input jitter. The other device, on the other hand, reacts directly to the input signal, whereby the actual mode of operation is part of the manufacturer's know-how.

Irrespective of this, it is clear that the reaction time can be reduced to a range of a few microseconds and below by relocating the high-speed signal processing to separate hardware. Additionally, any number of these applications can be installed in a system without any expected repercussions on the dynamics of the overall system. As a result, the skew position can be controlled directly on an intelligent field device and also several times in the machine. For example, a field device could provide this information with a high-resolution time stamp each time a light sensor signal changes or the time differences between $L_1$ and $L_2$ as well as $L_2$ and $L_3$ could be determined independently. The data is then sent to the CPU without any special measures, where it can be evaluated in a non-time-critical program section. Both solutions would be acceptable and economically justifiable even in the smallest systems.

As tempting as this solution is, it does harbor some risks. Manufacturers only offer intelligent field devices if there is a correspondingly high demand, and the devices can be sold in economically viable quantities. As a result, the implemented functions must largely meet the requirements of an industry and/or technology. The devices are therefore only equipped with precisely those functions in their hardware and software for which the widest possible acceptance is expected on the market. As a result, manufacturers will only offer new devices for a possible technology trend if the probability of market penetration is given to the required extent.

However, this means that these devices are also available to all other market players, meaning that the opportunities for a machine manufacturer to create unique selling points dwindle. In order to counter this weakness, the integrated functions are provided with as many options as possible so that there appears to be a possible variant for every application, and clever parameterization can generate its own advantages. For example, a commercially available frequency inverter sometimes has more than a hundred parameters in numerous function models for the optimum drive of pumps, fans, compressors, etc. The catch here is that a great deal of effort is required for correct parameterization, and sometimes there are only a few experts who have mastered this down to the last detail.

And what if a brand-new technology or the creation of modules requires special functions for which there are no intelligent field devices yet? It is almost impossible to persuade a manufacturer to make customer-specific software changes to an existing product in order to obtain a new functionality. After all, it is precisely for the reasons mentioned above that all other customers expect the greatest possible consistency.

And what if the functionality of a device is changed due to technical progress, making it incompatible with its predecessor model? How will existing machines and systems be serviced in that case? Or what happens if a device is discontinued by the manufacturer?

In conclusion, it can be said that intelligent field devices can provide highly dynamic functionality in a decentralized manner very quickly and at a reasonable cost. However, beyond special parameterization, the user has almost no influence on how to incorporate his own know-how into these devices.

✓ **Intelligent field devices**

+ Response times of less than 1 ms are possible depending on system performance and field device
+ Any high-speed I/O scope can be realized
− The functionality of the field device is developed for special requirement profiles for which there is the highest possible demand on the market
− Devices are manufactured in large quantities and are available to all market participants
− Sometimes complex parameterization generates higher engineering costs
− Customized individual software changes are almost impossible

### 5.2.3.4 *Special developments*

The basic structure and functionality of devices specially developed for a specific application correspond exactly to Figures 5.17 and 5.18 and the criteria of intelligent field devices. However, there is one major difference to consider.

While intelligent field devices are generally available to the entire market, this is not the case for devices developed individually and for a single customer. There are numerous companies around the world that tailor functions exactly to their customers' needs. A machine manufacturer's own electronics production facility may also be able to do this and is usually established for no other reason. However, suppliers of customized solutions must also meet the requirements for delivery reliability, certifications, and availability, and if all this is given, the question of cost-effectiveness remains. Even customer-specific devices can only be manufactured cost-effectively if a certain minimum number of units can be produced so that development costs, sometimes in the six-figure range, can be refinanced.

Once a customized device has been developed, it creates excellent unique selling points for the intended product. So, is this approach the solution?

Not at all because, on the one hand, the resources for in-house developments are usually not available or the economic conditions are not right. This is only possible for manufacturers who sell their machines in large quantities and can therefore shoulder the development costs. A special development for *skew* detection alone would certainly not be economically justifiable.

✓ **Special developments**

+ Response times of less than 1 ms are possible depending on the system performance and properties of the special devices
+ Any high-speed I/O scope with individual functionality can be realized
− Hardware and software are developed specifically for the device and the requirements
− Devices are manufactured in very small quantities, resulting in very high unit costs
− Delivery times, service and loyalty to the supplier must be considered
− Usually very high costs for changes

The question therefore remains as to how to implement dynamically sophisticated, individual functions within a module, but also in any other automation structures, without sacrificing flexibility and cost-effectiveness. A suitable solution approach is presented in the following section.

### 5.2.3.5 *Intelligent I/O modules*

It has always been a priority of manufacturers of automation systems to achieve even shorter reaction times. In the early 1990s, some manufacturers began to equip their systems with so-called *peripheral processors*. These devices were simply equipped with corresponding inputs and outputs and plugged into control systems like normal I/O modules.[19] They were used to implement drive functions, fast I/O operations, or in combination with corresponding HMI devices to implement sophisticated operating and visualization functions. The advantage of this technology is the possibility of mostly free and individual programming.

At the same time, I/O modules were equipped with increasingly powerful processors and additional functions. Examples of this are counting modules that can detect and evaluate highly dynamic signals and are now part of the standard portfolio of most manufacturers. The behavior and operation of these modules are similar to that of intelligent field devices. This means that they work with a fixed firmware and the exact functionality

---

[19]Examples include the WF723 positioning modules from Siemens and the PP40/60 peripheral processors from B&R.

is defined by a more or less complex parameterization. At the end of the 1990s, however, there were also I/O modules with firmware that could be created by the user [5]. Their user software was created either by using standard programming languages or modules from a special function library in the same engineering tool that was used to develop the rest of the application.

The advantage of this technology is that the entire high-speed application can be developed and, above all, maintained freely and without restrictions by the user. The know-how remains with the user, as functional enhancements can be made at any time under the user's own control. In addition, the program is developed precisely for the desired function, so there is no need for complex parameterization or only a few adjustments. It is therefore only logical that this technology should also be favored for the creation of modular concepts.

References [6, 7] display how reaction times of up to 1 $\mu$s can be achieved by using intelligent input and output modules that are freely programmable.[20] It is not necessarily the one microsecond but simply the possibility of achieving reaction times that are decoupled from the CPU and field bus in any automation structure that is of decisive importance for many high-speed applications.

Figure 5.20 shows the elements of the reaction time, and Figure 5.21 shows the basic signal curve in a decentralized structure.

The I/O module has optimized input and output physics for fast signal acquisition and output. The input signals are sampled by the I/O module, converted into logic signals, and then processed in the integrated processor. The resulting output is sent directly to the actuator via the output driver. This way, the I/O module takes over the functions of the central controller, which is thus relieved. In principle, the CPU can also be made smaller and reaction time of a few microseconds can still be achieved with a control CPU that operates in the millisecond range.

The user programs the respective technological functions, just like everything else, directly in the engineering software itself, which was presented in [6] with the *reACTION technology*. This is done in the form of function blocks in accordance with IEC 61131 using logical operations such as AND, OR, or XOR as well as arithmetic operations, such as ADD,

---

[20] In the articles mentioned, the input jitter is not included in the calculation of the reaction time.

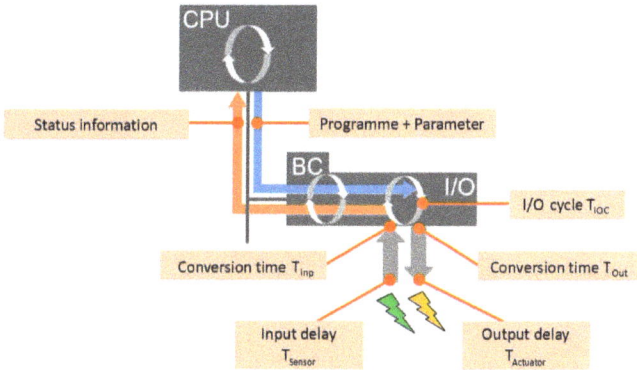

**Figure 5.20.** Elements of the reaction time with intelligent I/O modules.

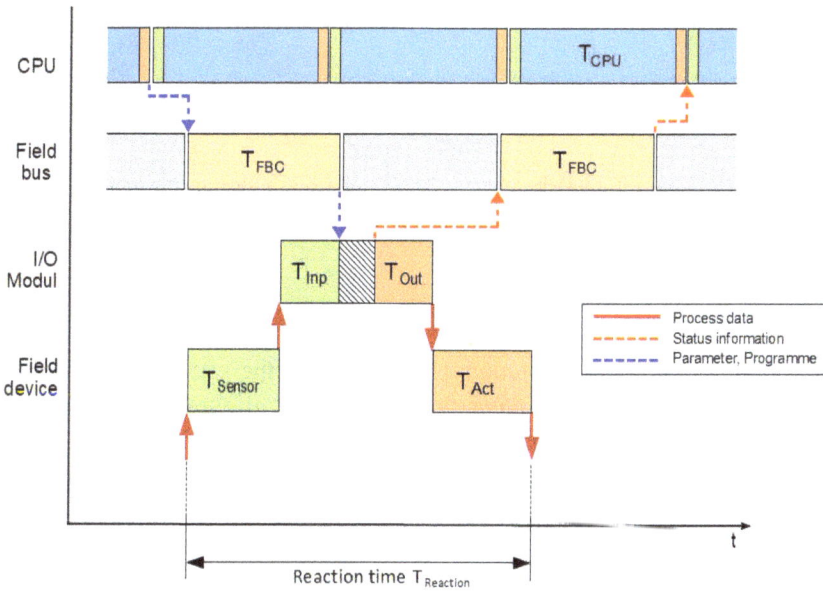

**Figure 5.21.** Principal signal curve in a decentralized structure with intelligent I/O modules.

PWM, comparator, or counter functions (Figure 5.22). This means that even more complex tasks such as sophisticated controllers can be realized. The circuits created using function blocks can be tested like classic control code by executing the modules on the controller. If everything works

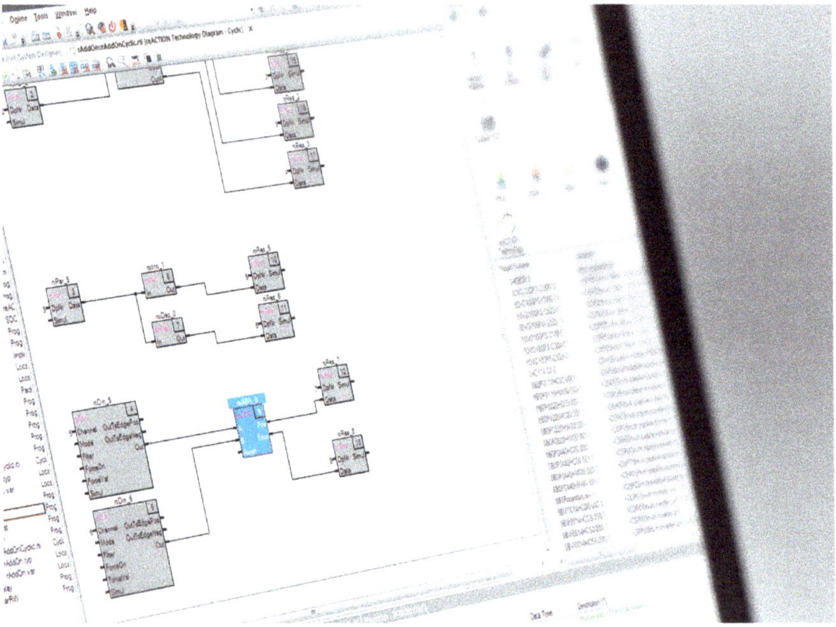

**Figure 5.22.**    The *reACTION* functions are programmed in a function diagram with function blocks in accordance with IEC 61131.

*Source*: B&R Industrial Automation GmbH.

perfectly, the software function is assigned to the executing hardware component simply by adapting the hardware configuration and is programmed and active on the respective I/O module one mouse click later.

In addition, technology-dependent changes to the reACTION program can also be exchanged within a few cycles during operation simply by calling a method in the application program of the higher-level CPU.

A solution with intelligent I/O modules would certainly be the best choice for the *skew detection* function. This way, the output signal for a subsequent ejector could be set directly from the speed calculated in the module itself and the time difference between the signal changes of $L_2$ and $L_3$. The evaluation method can even be adapted according to the specific implementation in the overall machine if this function is carried out in an autonomous module. If, for example, an exact master position is available, the I/O module can calculate this directly with the signal changes of $L_2$ and $L_3$ in program version A. If this value is not available, the I/O module evaluates it directly. If this value is not available, a program version B additionally evaluates $L_I$ and calculates the skew position via the speed. This way, the number of modules and the commissioning effort can be easily reduced if both variants can also be detected automatically. The corresponding control configuration corresponds to the illustration in Figure 5.13.

> ☑ **Intelligent I/O modules**
>
> + Response times of a few microseconds can be achieved depending on system performance
> + Any high-speed I/O scope possible
> + Hardware based on standards, therefore only slightly higher unit costs
> + Programming is done individually for the required functionality
> + Software changes are possible at any time during the entire life cycle
> + Software changes are also possible during operation
> + Individual programs ensure unique selling points
> ! Field devices must be selected according to the dynamic requirements

## 5.2.4 Summary and examples of solutions

Influences on the reaction time in centralized and decentralized control concepts were analyzed in the explanations and problem areas were identified that need to be considered in a modular design. Furthermore, different scenarios were considered as to how these influences can be eliminated or their effects minimized. In the following, further functions of a *saddle stitcher* for the production of stitched brochures are used to demonstrate the design possibilities.

### 5.2.4.1 *Thickness measurement of saddle-stitched brochures*

In order to produce wire-stitched brochures such as magazines, manuals, or leaflets, the individual folded printed sheets are placed on top of each other in *saddle stitchers*[21] from the inside out, stitched together with wire staples, and trimmed in a *three-sided trimmer*.

In addition, data carriers, postcards, or sample packs of cosmetics, for example, can be inserted using an *inserter* or *product sample adhesive* or

---

[21]See examples in [15, 31, 32].

glued onto specific pages with the exact motif. The thickness of the bro-chure is measured before stapling to ensure that, first, all sheets are present and, second, that the product samples and their placement are precise. Figure 5.23 shows the basic diagram of a thickness measurement with a pressure roller and sensor element.

As many measured values as possible are recorded and compared with a previously taught-in profile, depending on the design, in order to recognize the completeness and exact position of the inserted fabric sample. Further processing only takes place if the measured profile is identical.

The challenge is to carry out as many measurements as possible, even with a maximum machine output of 18,000 products/h.[22] With a processing time of 200 milliseconds per product, a maximum length of 320 mm, and a required accuracy of 1 mm, this requires a scanning time of 0.625 ms. That is a task which can be solved very well with intelligent

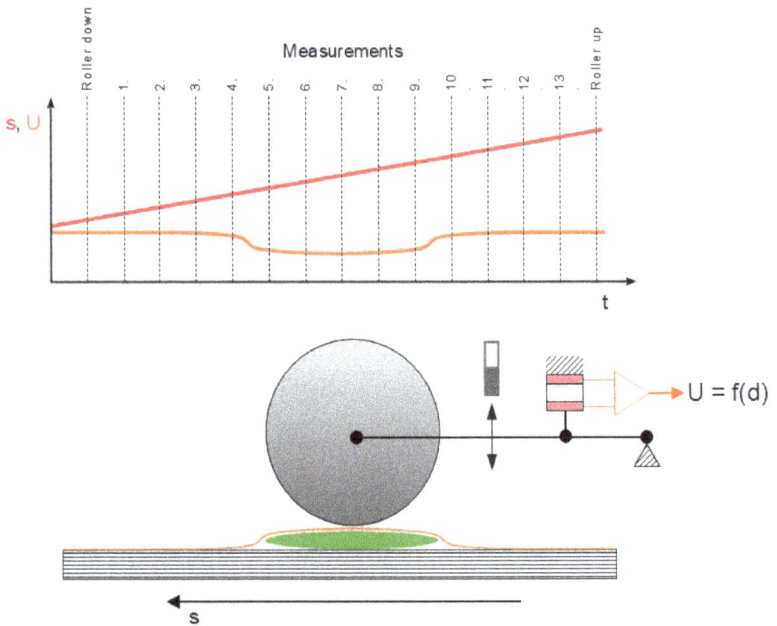

**Figure 5.23.** Schematic diagram of the thickness measurement function for a brochure with product sample.

---

[22]The following data refers to a real application as presented in [32].

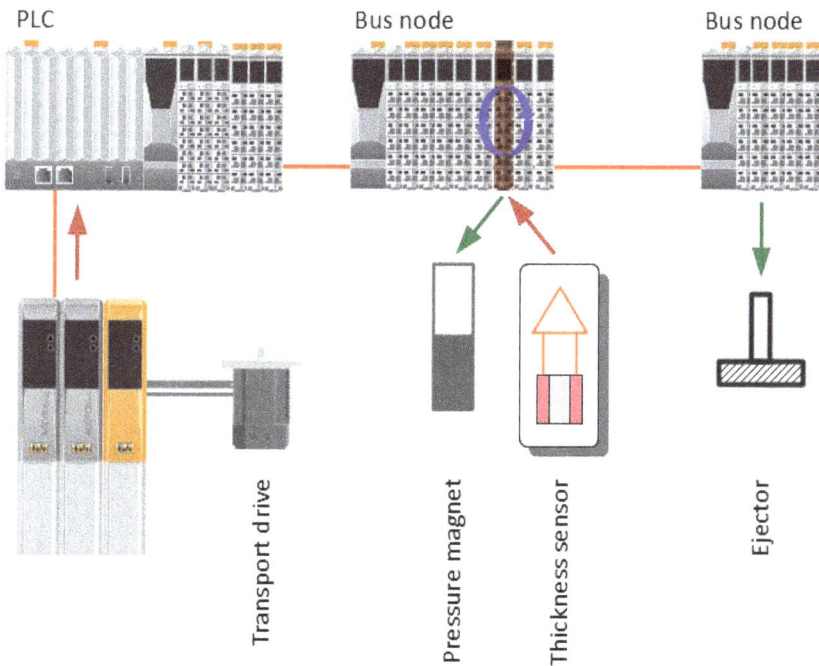

**Figure 5.24.** Automation diagram for the thickness measurement function in a saddle stitcher.

I/O modules. Figure 5.24 shows the automation diagram with two decentralized bus nodes and the servo drive of the transport system (here in the form of a collecting chain) around the stitching machine.[23]

The intelligent I/O module is located in one of the bus nodes and has an analog input for the measured thickness value and a digital output that is used to control a pressure magnet for the measuring roller. In order to place the pressure roller precisely on the front edge of a product and assign position information to each measured value, the module receives the synchronous position of the collection chain and therefore the exact position of the products via the field bus. The good/bad information is transferred to the control system which, if required, controls the exact

---

[23] The *stitching machine* is a part of the *saddle stitcher* in which the thickness control and subsequent stitching is carried out using wire staples.

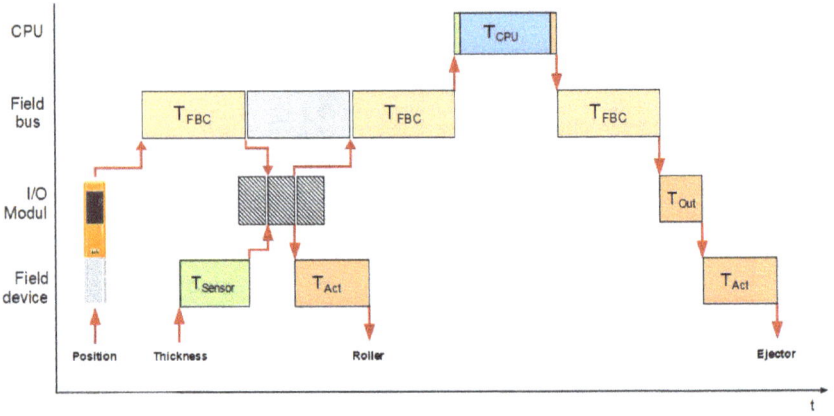

**Figure 5.25.** Signal curve for the thickness measurement function in a saddle stitcher with synchronous data transmission.

position of an ejector in the next cycle, which transports the faulty product into the waste container.

Figure 5.25 shows the signal curve and table 5.2 the relevant application values for this function. It is clear that their realization in the required dynamics with an intelligent I/O module does not pose a problem at all.

### 5.2.4.2 *Missing sheet control*

Depending on the manufacturer and equipment, a modern *saddle stitcher* can be equipped with up to 31 so-called *feeder modules* [8]. This means that *folded sheets* are removed from a stack at 31 positions and thrown onto a *gathering chain* one after the other. However, this also means that there can be 31 possibilities of a faulty print run, as it occurs more or less frequently due to gluing in the sheet stacks, for example. In extreme cases, if the first sheet does not land on the gathering chain without errors, this error could only be detected during the thickness measurement without a suitable check. However, this generates a large proportion of discards, which should of course be avoided.

To counter this, the feeder modules of the saddle stitcher presented in [8] can be equipped with a *missing sheet control* function. For this purpose, a camera is mounted on each feeder module in front of the gathering chain and a signal from the respective feeder triggers the image recording.

**Table 5.2.** Application values of the thickness measurement function with intelligent I/O module.

| Description | | Value of application | Remark |
|---|---|---|---|
| Input delay sensor | $T_{Sensor}$ | 50 $\mu s$ | – Measured value |
| Output delay pressure magnet | $T_{Press.mag.}$ | 120 ms | – Activation of the pressure solenoid with dynamic dead time compensated cam switch unit |
| Cycle time I/O-Modul | $T_{IOC}$ | 2 $\mu s$ | – Measured value<br>– Time for measured value acquisition, target/actual value comparison and calculation cam switch unit |
| Cycle time field bus | $T_{FBC}$ | 400 $\mu s$ | – Data transmission between the bus devices in real time |
| Response time measuring cycle | $T_{FBC} + T_{IOC}$ | 402 $\mu s$ | <<625 $\mu s$ |
| CPU cycle time | $T_{CPU}$ | 10 ms | – Cyclical processing of user applications |
| Output delay ejector | $T_{Ejector}$ | 180 ms | – Control of the ejector via a pneumatic cylinder |
| Ejector response time | $T_{CPU} + 3 \cdot T_{FBC} + T_{Ejector}$ | 191.2 ms | <200 ms |

The camera's image processing system then compares the captured image with a previously taught-in reference image. If there is a discrepancy, information is sent to the following feeders, which in turn no longer place a sheet.

In this application, the task is to trigger the image recording at the exact moment of placement and before acceleration by the collection chain. If this happens just a few microseconds too early or too late, the image would be blurred and therefore unusable due to the end of the drop or the start of the transport movement. The automation diagram of the *autonomous feeder module* is shown in Figure 5.26.

A special light sensor detects the falling sheet and sends this signal to an intelligent I/O module. The exact time for image acquisition is calculated and the command for triggering is transmitted based on the remaining drop height and the speed of the collection chain calculated in the previous I/O cycle. The evaluation result is then provided by the camera via the field bus as status information for the CPU and subsequent feeder modules.

The reaction time results from

$$T_{\text{Reaction}} \leq T_{\text{Sensor}} + T_{\text{Inp}} + T_{\text{IOZ}} + T_{\text{Out}} + T_{\text{Aktor}} \leq 3,2\ \mu s \qquad (5.11)$$

and is therefore even smaller than the technological specification of $T_{\text{Reaction}} \leq 4\ \mu s$.

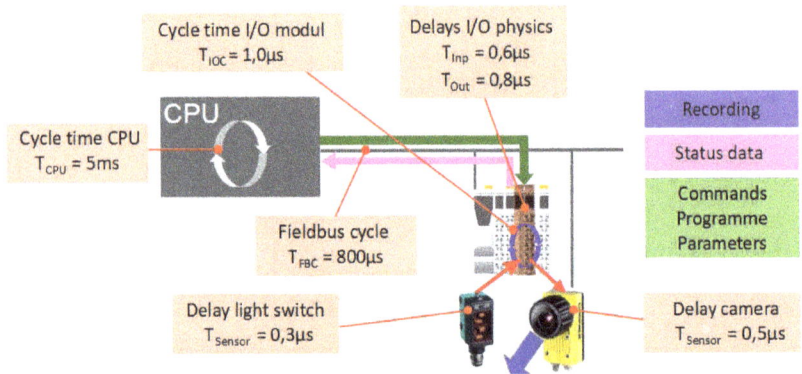

**Figure 5.26.** Automation diagram of a missing sheet check with intelligent I/O module, light scanner, and camera.

# 5.3 Machine safety

It has already been discussed in Chapter 1 that it is vital to ensure the safety of operators but also to protect production facilities, the environment, and the products themselves.

If a production system is modularized, further requirements are added because a modular system is by definition a structurally variable system. However, this is precisely a feature that is not expedient in *safety engineering*, as the manufacturer of a machine must ensure that a risk assessment is carried out for all options and that compliance with the applicable health and safety requirements is documented with a so-called *declaration of conformity*.[24] Consequently, if an individualized machine is offered in ten variants, then a correspondingly large number of test procedures must be carried out — a requirement that is diametrically opposed to the design principles of a modularized structure.

Manufacturers of safe automation technology are well aware of this fact, and there are numerous concepts to achieve an acceptable engineering effort even for modular systems with safety functionality. This section presents and evaluates the most important strategies without discussing the relevant standards and regulations in detail. A large amount of further literature is available for this purpose.[25] However, we cannot avoid a brief look at the *Machinery Directive* and its consequences for the design of safe automation.

## 5.3.1 Application of the machine directive in modular systems

The *Machinery Directive* has existed since 1995 and has been binding in Europe as Directive 2006/42/EC of the European Parliament and of the Council of May 17, 2006, on machinery and amending Directive 95/16/EC (recast), or Machinery Directive 2006/42/EC for short, since December 29, 2009. It describes uniform requirements for health and safety in the interaction between human and machine. Irrespective of the place and date of manufacture, all machines first used in the European Economic Area from January 1, 1995, are subject to the EU Machinery Directive and must

---

[24]Cf. [9], Annex 1, "General principles", para. 1.
[25]The online *safety compendium* from Pilz GmbH & Co. KG [16] is suitable for a comprehensive introduction to the topic of machinery safety.

therefore bear a CE-marking. Accordingly, if a device, machine, or system bears this mark, it is ensured that a machine complies with the safety regulations, and by issuing an EC Declaration of Conformity, the manufacturer declares that he has taken all directives applicable to the product into account [9].

Numerous standards are listed under the Machinery Directive, starting with the general design principles for safety-related parts of control systems and their validation (EN ISO 13849 — 1 and — 2) [1, 10]. They also define rules for safety-related electrical, electronic, and programmable electronic control systems (EN 61508) [11] through to general rules and profile specifications for functionally safe transmission in field buses in industrial communication networks (EN 61784 — 3) [12] and for the manufacture and operation of safe machinery and equipment [13].

In principle, the Machinery Directive describes requirements for new machinery. However, if a machine is modified, it must be ensured that no new hazards arise. To this end, it must be clarified whether the modification results in so-called significant changes, and if this is the case, the machine must undergo the same procedures as during its manufacture. In principle, Figure 5.27 shows how this should be done. It is therefore obvious that no new hazards or risks should arise when adding or removing any machine modules.[26]

Furthermore, the same certification process applies to a modular machine, which consists of several assembled individual machines, as for production lines. Each individual system must meet the machine safety requirements for itself and for the system as a whole. After all, an event on one machine or module can have a safety-related effect on another machine, the entire production line, or, in modular designs, the rest of the machine. Each subsystem therefore also bears a CE marking in its own right, just as the entire system must also undergo a CE marking process.

This problem can be easily illustrated using a *saddle stitcher* in which the individual sheet feeders are designed as autonomous modules [15]. Just like the rest of the machines, these are each individually safety-tested and validated. However, if an individual feeder is removed, there is an empty space where the exposed transport chain creates an additional

---

[26]The BG RCI information paper on *Modification of machinery* [44] is based on the specifications of the German BMAS from 2015 [14] and explicitly includes the use of control technology in the version updated in 2020. To this end, the application of the illustrated graph is explained in numerous case studies.

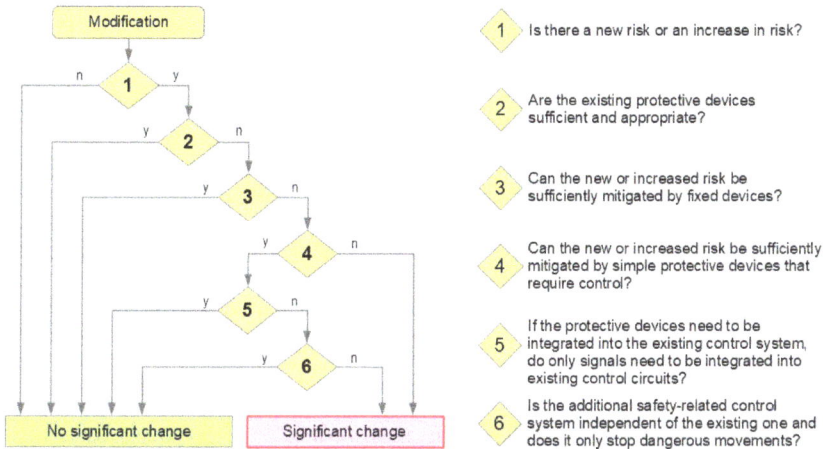

**Figure 5.27.**   Decision diagram "Substantial modification of machinery".

*Source*: According to [44].

hazard. Therefore, it must be prevented that a module can be removed while the machine is running. Furthermore, after removing the feeder, the operator must eliminate this empty space either by using another feeder or by attaching a cover, and it must be ensured that no dangerous movement of the transport chain can occur during this conversion. In the saddle stitcher described in [15], each module position is monitored by a safety switch that is actuated either by the feeder or a cover. If it is not, the safety controller in the safe servo drive of the transport chain activates the safe motion function *Safe Operational Stop* (SOS), and to prevent any motivation for manipulation, the cover is attached to a hinge next to the transport chain and only needs to be folded up with one hand movement. This means that the entire changeover process takes less than a minute, which is a unique selling point for this saddle stitcher.

This case study produces the graph shown in Figure 5.28.

So far, so good. But what does this mean for automation technology? From the moment the feeder is added to or removed from the saddle stitcher, there is at least one emergency stop switch more or less in the entire machine — the one or ones of the individual feeder. However, it is stipulated that each emergency stop switch must act on the area of a machine that can be seen from the point at which it is mounted. This can also apply to the entire machine or production line. After all, an operator standing at one end of the machine and recognizing a hazard at the other

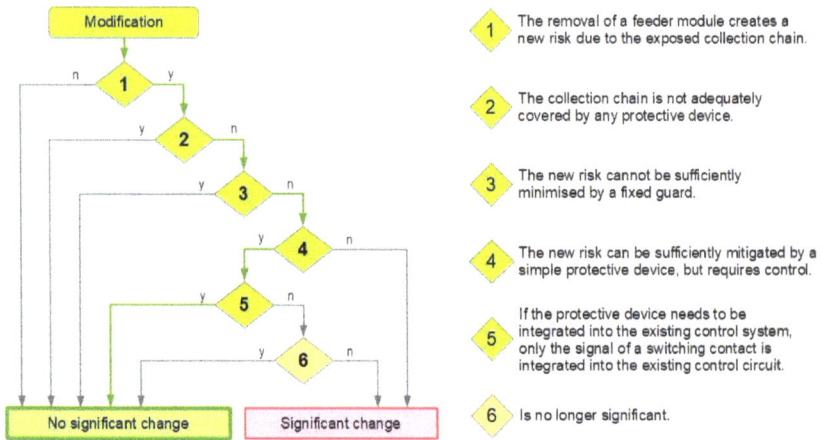

**Figure 5.28.**    Decision graph for removing a feeder module in a saddle stitcher.

end must be able to intervene immediately. He should not have to think first about which emergency stop switch is intended for this or that area. This would result in valuable time being lost, which could potentially cost the life of the person at risk.

As a result, it must be ensured that the safety-related part of the machine control system recognizes the removed or added safety devices as automatically as possible when the module configuration is changed and handles them accordingly. The solution used in [15] requires the operator to enter a brief dialog at the main control panel after the new components have been detected automatically, and after a few seconds, the machine is ready for operation and, above all, remains safe.

For automation, this requirement means that a module must be capable of safety-related communication in addition to the purely technologically oriented operational data traffic. In the next section, we look at the different variants of safe switching and automation technology. But first, we want to get an overview of established safety-related concepts.

### 5.3.2  Safety technology at a glance

From the very beginning, when only an electric motor was used as the main drive in the machines, there had to be a device with which the machine could be brought to a standstill in the event of imminent danger. The emergency stop loop introduced at that time was very simple and straightforward. One or more opening switches were simply connected to the holding circuit of the main contactor, thus disconnecting the motor from the power supply.

However, it is no longer as easy as it is shown in Figure 5.29. For example, *EN ISO 13850:2008 — Safety of machinery — Emergency stop — Design* principles stipulates that the supply lines to the *emergency stop switches* must be tested continuously against earth faults and short circuits. In addition, it gradually became clear that this type of safety system could be disabled in the event of a fault and the protective function would therefore no longer be provided [16].

Since this is not possible in this circuit, there are special *emergency stop devices* that are specially certified for the various safety requirements[27] (Figure 5.30).

Furthermore, there are special switching devices for many standard technological applications, such as presses or bending machines, which

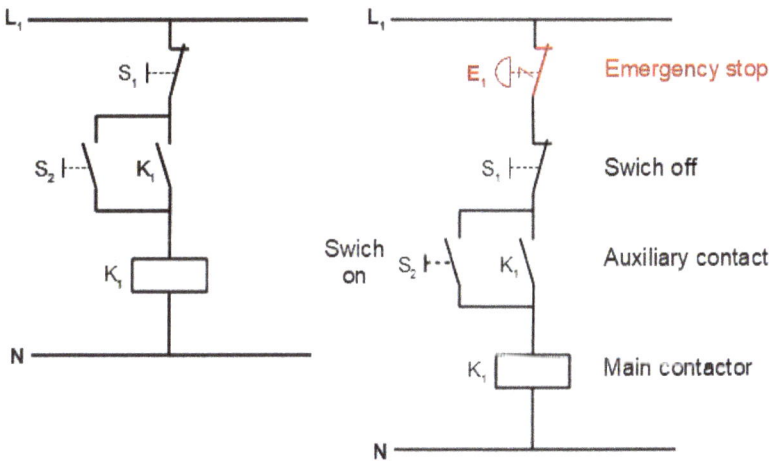

**Figure 5.29.** The principle of a simple contactor circuit with an additional emergency stop switch has not been permissible for a long time.

---

[27] See corresponding products in [33, 34] and many more.

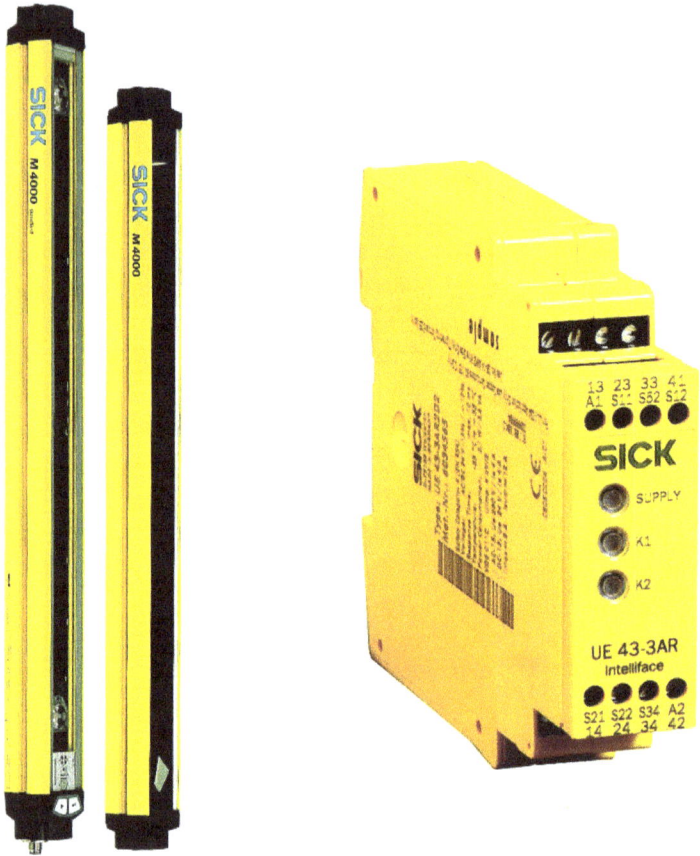

**Figure 5.30.** Safety light barrier and switching device for monitoring protected areas.
*Source*: Sick AG.

realize the necessary safety functions in combination with safe field devices (safety light curtains, speed monitors, laser scanners, etc.). Figure 5.31 shows the variety of possible safety devices using the example of a palletizing system.

The enormous increase in the quality and quantity of safety requirements and the amount of device technology required as a result of the *Machinery Directive* alone has meant that the realization of the necessary electrical design can become quite confusing.

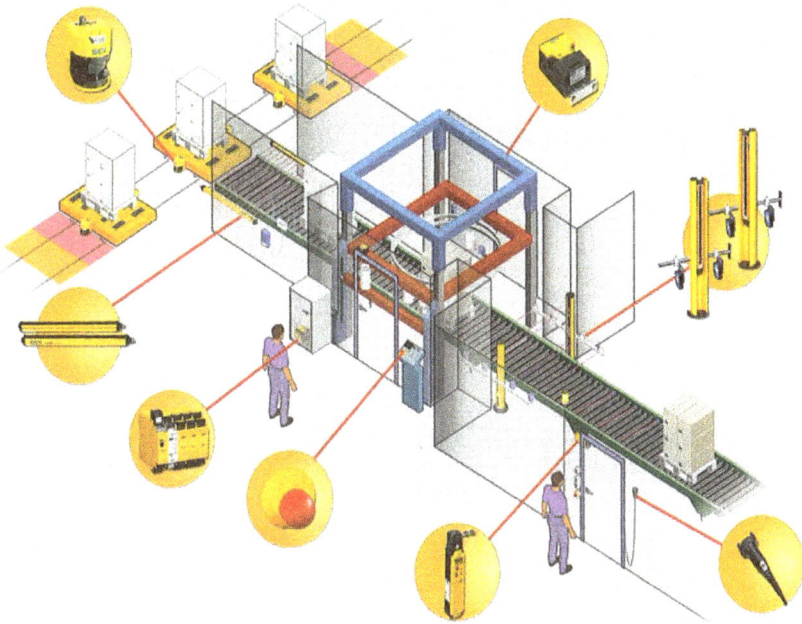

**Figure 5.31.** Principal representation of the safety technology on a palletizing system.
*Source*: Sick AG.

Figure 5.32 shows a relatively simple application with two emergency stop switches, an operating mode selector switch, a light curtain, and a speed monitor for a *safely limited speed*, as it was the state of the art just a few years ago. The connections shown correspond approximately to reality and demonstrate the enormous amount of wiring required. Just imagine having to replace the safety relay block — certainly a task, which is time-consuming and prone to errors.

In addition, there are parameters in the individual devices that all have to be set and checked individually and in some cases with special diagnostic devices.

Figure 5.33 shows how the palletizing system can be equipped with dedicated safety switching devices. The individual devices are located in a central control cabinet or in decentralized control boxes at technologically and structurally suitable locations.

Fortunately, things are better now. Thanks to modern electronic safety switching devices, wiring, diagnostics, and the entire engineering in a complex safety application can be significantly simplified. For this

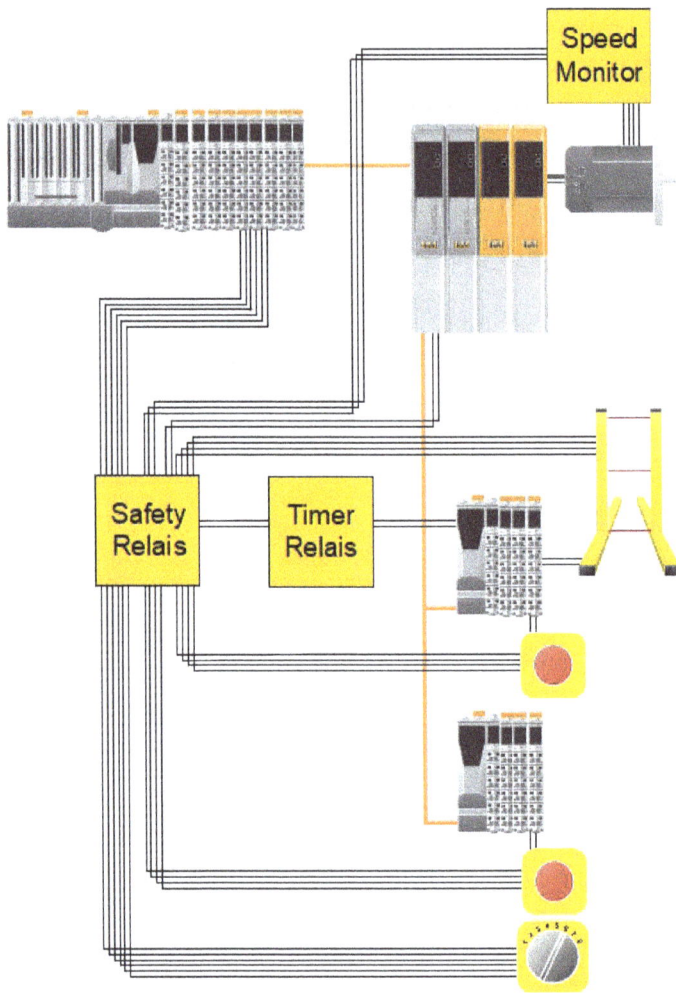

**Figure 5.32.**    Example of a safety application with dedicated safety components.

purpose, the safety switching devices are equipped with communication interfaces via which all non-safe information (diagnostic and status data, parameters, etc.) can be transmitted to the central control system (Figure 5.34). For example, the machine control system can be informed and displayed on HMI devices which emergency stop switch has been triggered or which safe operating mode has been set by an operating mode selector switch.

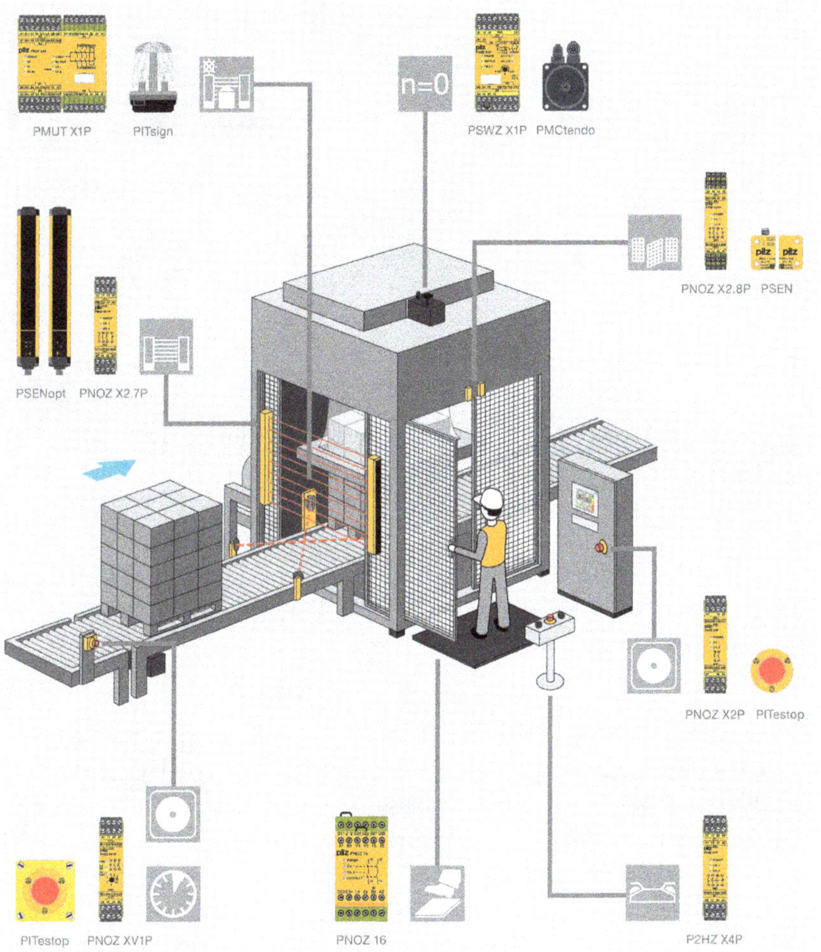

**Figure 5.33.**   Safety application of a palletizing system with safety relays.
*Source*: Pilz GmbH & Co. KG [36].

If this system is now to be modularized, regardless of whether it is based on the integrated or autonomous principle, the designer again has to deal with the problem of cross-modular safety functions which were discussed in the previous section. It must therefore be possible for the safety components to also exchange safety-relevant data via a suitable *safety bus*. This enables the control systems to react to any safety-relevant events and information from the entire production system as required.

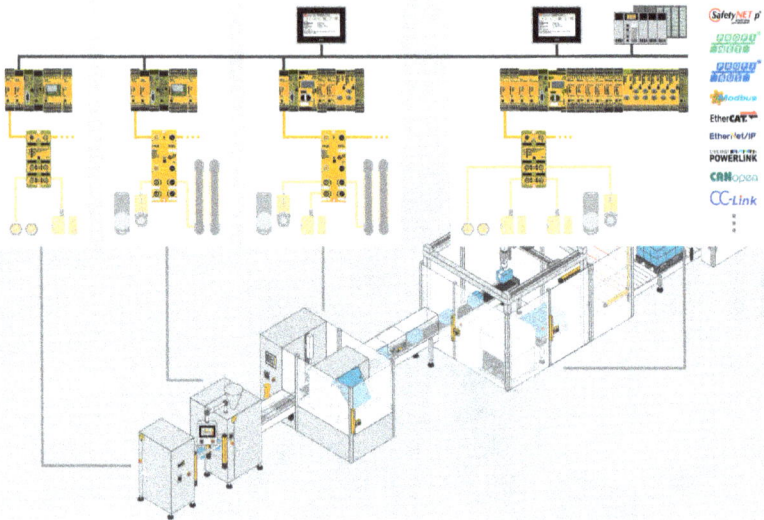

**Figure 5.34.**    Safety application with communication-capable safety relays.
*Source*: Pilz GmbH & Co. KG [36].

Figure 5.35 uses the example of a modular palletizing system to show how the safety functions with *distributed Safety controllers* can be realized. Figure 5.35 reveals another detail. As we know, there is not only safety-relevant functionality in a production system, even if its proportion can vary greatly from case to case. Therefore, in most modern *safety control systems*, it is also possible to process purely functional applications. In this case, the terms "Safety-integrated control system" or "*integrated safety technology*" are used, depending on which side it is viewed from. In Figure 5.35, this is symbolized by the gray I/O modules[28] in the individual control racks. This means that these control systems can be configured in the same modular way for the respective requirements as their "gray" counterparts, and complete production systems and lines can be configured with these safety controls. Within the following section, we discuss whether and under which conditions this makes sense and which alternative solutions are available.

---

[28] The terms "gray" and "yellow" are used by many manufacturers as synonyms to differentiate between functional (gray) and safety (yellow) application.

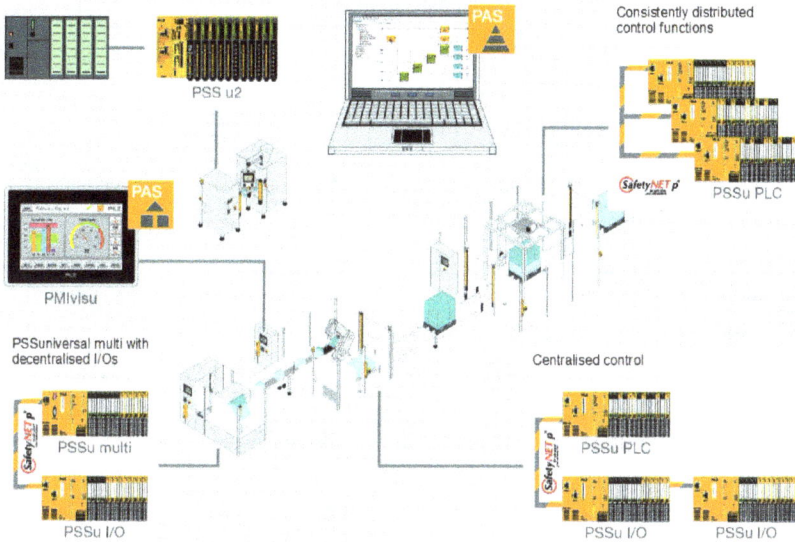

**Figure 5.35.**   Modular safety application with safety controllers.

*Source*: Pilz GmbH & Co. KG [36].

## 5.3.3 Safety control technology

The previous explanations have shown the path from the simple emergency stop via *safety switching devices* to *safe control technology*. Consequently, safety aspects have always been and continue to be at the forefront of this development. Safety systems are therefore designed in a way in which no additional hazards are generated by errors, whether in hardware or software, and the safe state of a machine is never left[29] as far as possible. This is achieved through a whole range of constructive measures.

Safety-related electrical devices, whether *safety switching devices*, *safety control units*, or their peripherals, are developed and validated in accordance with EN 61508 or EN ISO 13849 and many others. In control systems and switching devices, this begins with the connection of the sensors and actuators. In order to ensure that the connection has been

---

[29]"As far as possible" in this context means that the machine or system must be equipped with the necessary safety technology at the time it is placed on the market in accordance with the "State of the art in science and technology" [16].

**Table 5.3.**    Fault detection through internal pulse generation of a safe input module [17].

| Description | Error during contact | |
|---|---|---|
| | **Open** | **Closed** |
| Ground fault on pulse output | is recognized | is recognized |
| Switch off against 24 VDC on pulse output | is recognized | is recognized |
| Cross-circuit between pulse output and other pulse signals | is recognized | is recognized |
| Ground fault on signal input | **is not recognized** | is recognized |
| Switch off against 24 VDC on signal input | is recognized | is recognized |
| Cross-circuit between signal input and other pulse signals | is recognized | is recognized |
| Cross-circuit between pulse output and signal input | is recognized | **is not recognized** |
| Wire breakage | **is not recognized** | is recognized |

implemented without errors and is permanently functional during operation, continuous tests must be carried out during operation, the scope of which must be determined in a risk analysis. Therefore, most manufacturers equip their safe digital I/O modules with pulse outputs, which can be used to detect wiring problems such as short-circuits or earth faults, short-circuits against 24 VDC, or cross-circuits[30] (Table 5.3).

However, this only tests the electrical connection, which does not reveal anything about the functionality of the connected field devices. While intelligent devices can diagnose themselves and signal that they have left the safe state, this is not possible with passive devices such as emergency stop buttons (Figure 5.36).

One option for monitoring the function of these devices is the two-channel design of the input circuits. Figure 5.37 shows the circuit and evaluation for an emergency stop button in an application for which a *Performance Level e*[31] is to be implemented.

---

[30]This is also done in the safety switching devices.

[31]See Section 3.4 in [16] "Determining the required performance level $PL_R$".

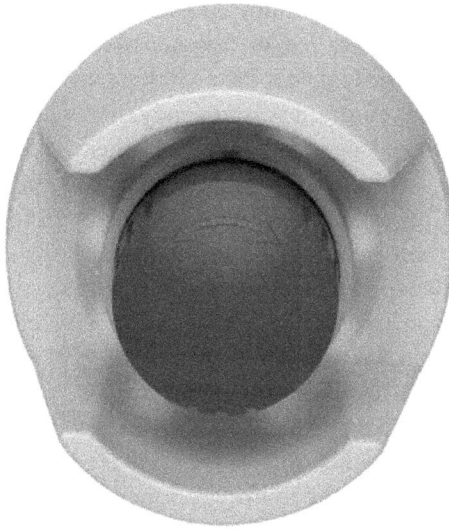

**Figure 5.36.**   Emergency stop button IP65, illuminated with protective collar.
*Source*: Pilz GmbH & Co KG [35].

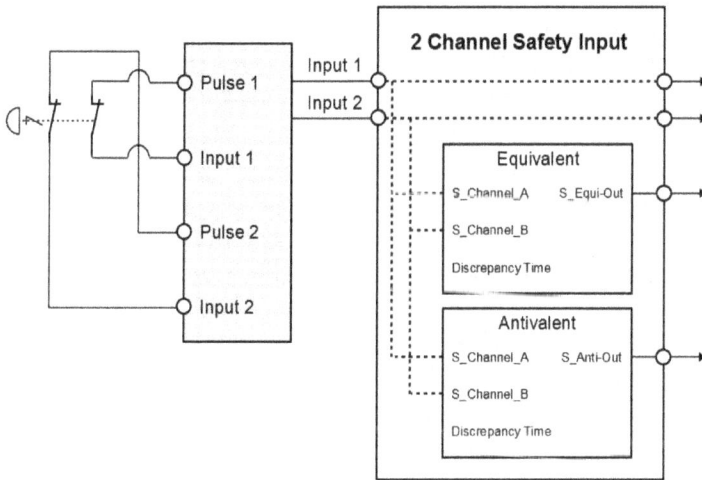

**Figure 5.37.**   Alternatively, the normally closed/normally open combination can also be used. In this case, the evaluation is carried out with an antivalence check.

In the case shown, a control device is used that has two separate NC contacts,[32] which are connected to two separate inputs of a safe input module. In order to reliably detect the switching state, both contacts must change state within an adjustable discrepancy time when actuated. Only if this change is recognized as valid it may be further processed by the safety controller. If this is not the case, a fault is suspected and the system must be shut down with the fastest possible reaction time and thus transferred to a safe state. For this purpose, the I/O module shown in the figure has an internal equivalence evaluation for contacts that switch in the same direction as shown in Figure 5.37 and an anti-equivalence evaluation for contacts that switch in the opposite direction.

However, the CPU of the safety controller, its firmware, the internal communication, and the controllers on each individual module must also meet these requirements. For this purpose, a procedure is used that we know from everyday life as the four-eyes principle, in which, for example, an important contract is checked by another person. The contract is only signed if everyone agrees, otherwise the contract is not concluded. The approval principle in a safety controller works exactly according to this principle. According to this principle, a device only works faultlessly if two parallel and separate safety circuits come to the same conclusion. This principle also applies to the interaction of functional and non-safe "gray" and safe "yellow" applications. For example, a "gray" switching command can also be sent to a safe output. However, this is only executed if the "yellow" safety application agrees. Figure 5.38 shows this principle for a safe digital output in a concept with integrated Safety technology.

This also illustrates a basic principle of how safe and non-safe applications have to deal with each other. Quite simply, the "yellow" part of a safety controller has sovereign rights to all safe I/O channels. For its part, the functional "gray" application has the same access rights to all "gray" inputs and outputs. However, its horizons are broadened by the fact that it can also read the entire safety-related part at any time and in any state. It only requires the consent of the "yellow" application to be able to actively trigger safe switching commands.

It is also a major advantage of this technology that the "gray" side has read-only access to the entire system because no additional data exchange is required, as it would be the case if safety switching devices were used.

---

[32] Alternatively, the normally closed/normally open combination can also be used. In this case, the evaluation is carried out with an antivalence check.

The "gray" application controls safety I/O with consent of the safety application

&

The „gray" application controls „gray" I/O as usual

The „gray" application can all safety I/O read

The safety application controls safety I/O also alone

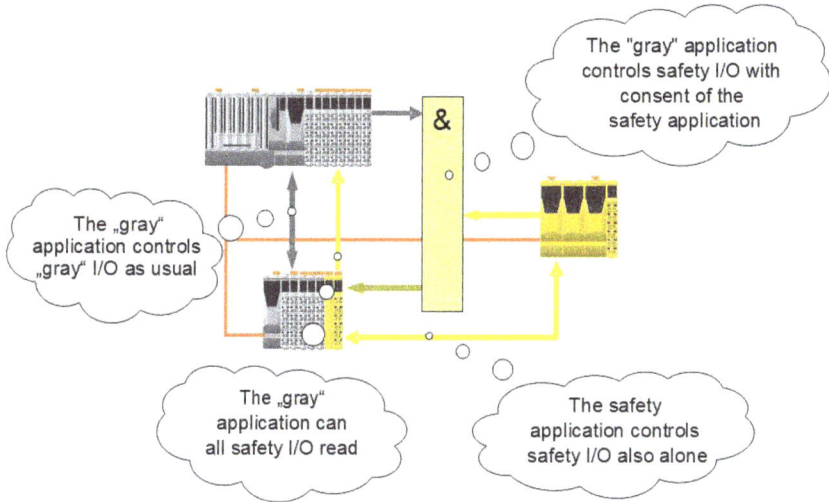

**Figure 5.38.** Enabling principle: Safe outputs can only be influenced by the functional application with the approval of the safety application.

If, for example, a specific operating mode is set via a safe operating mode selector switch, this information is sent directly to both sides, and while the "gray" application carries out all the intended status changes, it is monitored by the safe "yellow" part. Or there is direct interaction between the two applications, as can be easily demonstrated with the safe door locking system.

Let's take another look at Figure 5.33. The operator must register if he wants to enter the locked area in this system, for example, to make a setting. In the simplest case, this would be a button press on the HMI or a key switch. The sequence shown in Figure 5.39 starts when this safe signal change is triggered.

Access is only unlocked by the "yellow" control unit once the safety-relevant conditions have been met and the corresponding monitoring parameters have been set. For example, if the system can continue to operate at a safely reduced speed, the functional part must first reduce the speed of all enclosed and relevant drives to a maximum of this value. The "yellow" part is only informed when this has been achieved and, if necessary, other actuators have been set to the required status. Figure 5.40 shows this interaction using the example of *Safe Motion* and the requirement for a safely limited speed with a function named *Safe Limited Speed* (SLS).

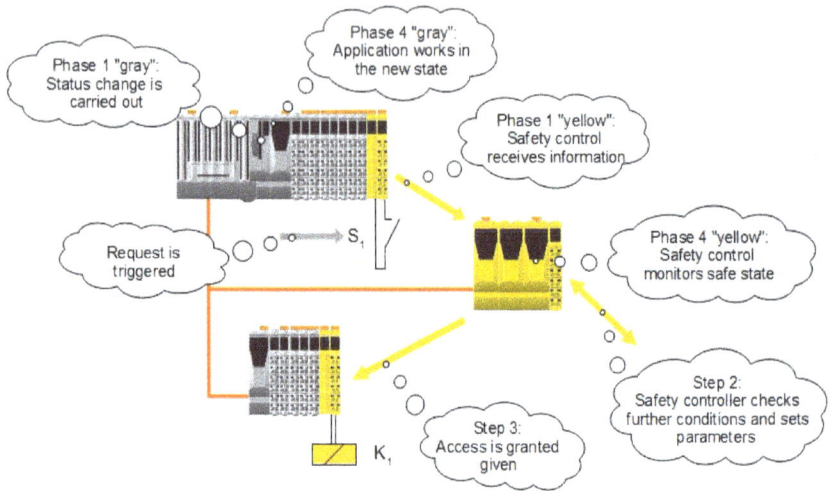

**Figure 5.39.**    Interaction of functional and safe application using the example of the Safe Motion function.

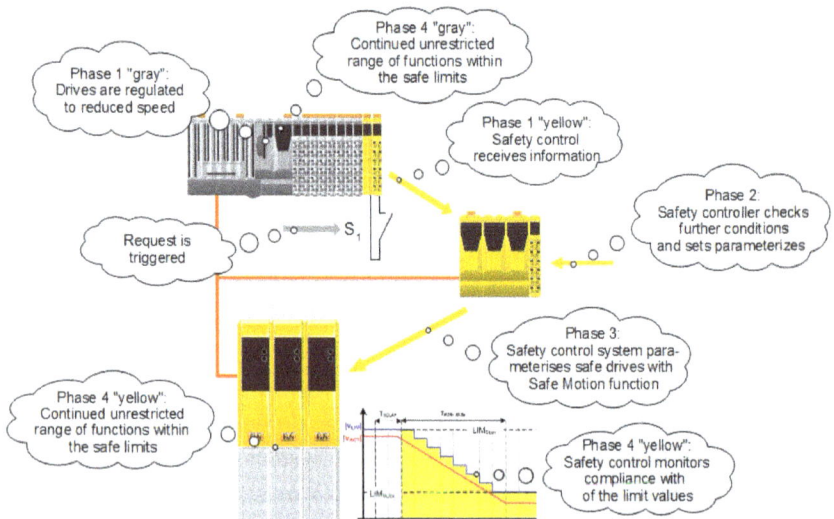

**Figure 5.40.**    Interaction of functional and safe application using the example of access control of a securely locked access door to the hazardous working area of a machine.

In this application, the enabling principle works in a way in which the safe "yellow" part of the control system generally permits all movements as long as the limit values of the activated safe motion function are not violated.[33] In this case, the *Safe Stop 1* (SS1) command would be triggered immediately by the safety controller and the safe restart interlock would then be activated with *Safe Torque Off* (STO).[34]

In the last pictures on this topic, an automation structure was used in which a functional "gray" and a safe "yellow" controller access common input and output modules coupled via a field bus. Figure 5.35, on the other hand, shows safety controllers that have been supplemented with "gray" I/O modules and follows a slightly different approach. Even if the interaction between the "gray" and "yellow" parts basically follows the same rules, there are still some differences that need to be considered, especially in a modular machine, which we want to work out in the following section.

### 5.3.4 Supplement safety technology or integrate them?

It is worth taking another brief look at the history of the development of safe control technology to better understand the problem. For a long time, safety switching devices dominated the market, which were configured for their intended use simply by setting operating modes and parameters via DIL switches or even with diagnostic devices and computers. In the beginning, communication with the functional part of the automation system was achieved solely via appropriate wiring to additional inputs and outputs of the "gray" controllers, but today safety components also have serial interfaces for direct data exchange. These devices are still being produced and have their right to exist in many applications.

However, the strict separation of the two hardware platforms is certainly becoming an obstacle to the rapidly increasing technological and safety requirements. Safety functions of increasing quality and quantity as well as new safety technologies (keyword: safe motion) are indispensable in order to achieve the objectives discussed in Chapter 1. In this constellation, however, their application leads to a sharp increase in complexity, which ultimately generates additional engineering effort that should be avoided.

---

[33] This also applies to coupled drives with asymmetrical movement profiles.
[34] See the different Safe Motion functions in [16], Section 7.4.

This trend was already recognized around the turn of the millennium by both the manufacturers of "yellow" safety technology and those of functional "gray" control technology, and both sides began to successively expand their systems to include the other part. As it was expected, more and more unique selling points have emerged for both approaches.

Which system should the designer choose, and which approach is better for object-oriented modularization? As is so often the case, there is no clear answer, as this also depends on the general conditions of the application, and it is therefore not possible to make a universally valid system comparison. Basically, both approaches are subject to the same safety-related laws, standards, and regulations, and both camps also have broad product portfolios available. One is a little more compact, the other a little faster, and the next a little cheaper.

The proportionate quantity of safety technology in an application is certainly a relevant decision criterion.

For example, if most functions in a pharmaceutical plant must be located on the "yellow" side due to high health risks and almost all hardware components are therefore safety modules, a safety controller will certainly be the first choice. If, on the other hand, only very few safety-relevant signals are processed in a circular knitting machine with only one sophisticated safe motion function, then the answer is quite different.

Another aspect could also be the reaction time to be realized for the individual functions because due to the amount of safety tests and plausibility checks, safety control systems are usually at a dynamic disadvantage. Although this is increasingly being compensated for by ever faster processors and optimized hardware and software, the other side does the same. Therefore, the dynamic characteristic values must be compared with the requirements before deciding in favor of a safety controller.

Furthermore, the possible range of functions of the firmware in combination with the engineering system can be decisive. Added to this is the range of ready-made technology packages for the safe control of presses, bending or plastics processing machines, and many other library functions. The separation of physics and logic recommended for successful modularization should also be considered. If, for example, it is necessary to delve deep into the programming bag of tricks for new technological approaches, this can be a knock-out criterion for the safety controller.

The reason for this is an understanding of the requirements for safe programming in accordance with EN ISO 13849, according to which

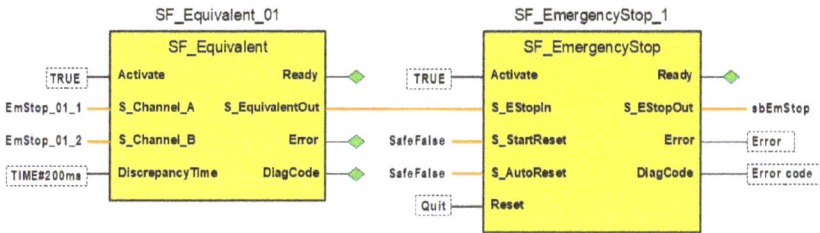

**Figure 5.41.** Functional diagram in the B&R Safe Designer for the evaluation of a 2-channel emergency stop button.

*Source*: B&R Industrial Automation GmbH.

validated function block libraries should be used whenever possible. Most manufacturers offer numerous function blocks which are already validated and certified in the libraries of their programming tools. It is also possible to use application-specific libraries that have been developed and documented in accordance with the requirements for the development of the *Safety Relate Application software* (SRASW) as per EN ISO 13849 — 1. Figure 5.41 shows such a program for the evaluation of a two-channel emergency stop button with safe *PLC open function blocks*.

In order to enable software developers, the greatest possible freedom in the choice of functions and commands in safety controllers, they allow a mixture of safety-related and non-safety-related programming, whereby these software parts are implemented in different function blocks and with defined interfaces. There cannot be any logical links between safety-related and non-safety-related data that would reduce the integrity of the safe signals. Programming languages with a limited range of languages or functions, so-called *Limited Variability Languages* (LVL), are offered in the development tools for creating the SRASW to prevent this. These languages are characterized by the ability to combine already developed library elements with new application code and thus comply with the required specification of the safety function. Classic examples of LVL languages are PLC languages such as *ladder diagrams* or *function charts*. In some systems, it is also possible to use the *Structured Text* programming language or high-level languages such as *C* or *C++*. However, their functionality is limited to such an extent that they can also be counted as

LVL languages.[35] Software developers are therefore limited on both sides, but this can be particularly painful in the functional "gray" part.[36]

Finally, the communication capability must also be taken into consideration when deciding on a system. Even safety controllers that do not have open yet safe interfaces and cannot operate open safety protocols will soon end up as electronic waste. But safety-integrated "gray" systems also need these communication options, and both are required to exchange safety-related data in the same way using protocols that are as open as possible. More on this in Section 5.4, where this topic is dealt with in more detail.

Which consequences does this have for the modular design of a machine? The different features of *integrated safety technology* can have a decisive impact on the type and nature of a modularization process. Autonomous modules can be created that are sensibly equipped with only one safety controller if these features are included in the analysis process. Other modules can perhaps be designed so that they can manage without safety technology because they do not pose any risk and the nearest emergency stop button is located in the immediate vicinity of the backbone. This means that the aforementioned selection criteria can be applied to each individual module for the automation technology of autonomous modules. Strictly according to the motto: the most suitable technology for each function.

As good as this approach may sound, it also has disadvantages. Caution is advised if automation systems for the different modules are purchased from different manufacturers who only offer one technology, i.e. only "gray" or only "yellow". In this case, it is first the number of different components that reduces the number of identical parts used and thus increases both manufacturing costs and service costs. In this case, there would also be different programming and diagnostic tools, which generate additional costs for acquisition and training. It is therefore better to choose a manufacturer that offers suitable systems for both sides.

Another point of attention concerns the life cycle of a modular machine. Let's assume that in the first draft, the decision was made for a singular, i.e. either a safety or a classic "gray" automation system.

---

[35] See Section 3.7 in [16].

[36] It should not go unmentioned that even developers of non-secure software with a defensive way of working produce more stable results than those who tend to celebrate their self-fulfillment in the complexity of their programs.

Now new functions are to be added in an update, for the effective process-ing of which the system to be expanded is not suitable. Then it could hap-pen that the elaborately applied safety control system has to make way after all, and the entire validation process becomes just as time-consuming as with the new development (which is what it ultimately boils down to).

If, on the other hand, a safety-integrated approach as shown in Figure 5.42 is chosen, this risk does not exist and new functions can be implemented on the respective side regardless of whether they are "gray" or "yellow" or a combination of the two. The validation of the safety func-tions must also be carried out in this case, but the overall effort should be significantly lower.

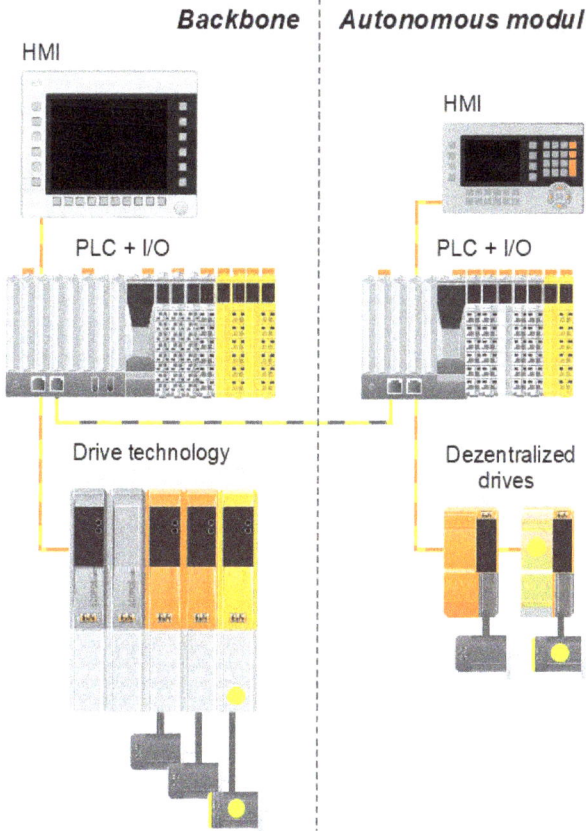

**Figure 5.42.** Modular system with integrated safety technology and secure communication.

So, if the backbone and the respective modules have their own control intelligence and corresponding communication options, as documented in [18], this approach combines the advantages of "gray" and "yellow" control technology in one system without restricting their possibilities and performance data. Furthermore, almost any extensions can be made on either side without the other world even noticing or being restricted. And finally, the programming tool has separate editors but manages the entire application in just one project. In addition, the system in [18] offers the advantage of managing *safety options* in just one project, which also significantly simplifies the modular design of a machine with integrable modules.

This leaves the topic of secure and non-secure communication, which we address in Section 5.4.

### 5.3.5  Summary

The protection of people, machines, the environment, and products against hazardous conditions in machines or systems cannot be ignored, even in their modular design. Therefore, automation concepts that can cover both functional and safe applications without restricting them must be selected. A prerequisite for successful modularization in addition to safe communication capability is the ability to freely design the control system required for the technological functions. This also means that the safe and non-safe hardware and software can be further developed without repercussions during the life cycle of a module. It has been shown that *safety-integrated automation technology* appears to be the most suitable for functional and object-oriented modularization.

## 5.4  Communication is (almost) everything

Not even 100 books would be enough to cover the topic of *digital communication* comprehensively because in our networked world, communication is no longer just *almost* everything. Whether it's digital television or listening to the radio, the telephone, the Internet, or the globalized world of work, nothing works without digital networking. The attentive reader will have noticed that the topic of *communication plays* an important, and in some cases, even plays a decisive role in every chapter of this book. After all, what would the *Industrial Internet of Things* (IIoT) or *digital production* be without fast and, above all, reliable data exchange? How would modern automation concepts with decentralized I/O modules, drive systems, and operating devices work without a real-time capable, fast field bus, and how is a modular system supposed to work at all without a data connection?

The topic of *communication* is so multifaceted that in this section we have to focus strongly on the specific topics that are important for the design of modular automation technology. We first take a historical approach to the various requirements for data traffic in the industrial environment, before briefly presenting and evaluating the most important field buses for automation. Finally, these findings are applied to the modular system in the digital production environment, and we do not forget the issue of data security.

## 5.4.1 Industrial communication at a glance

When Siemens and the German Federal Ministry of Education and Research initiated one of the first industrial field buses, the *Profibus* (Process Field Bus), in 1989, it seemed like a revolution. The goal of "implementing and disseminating a bit-serial field bus including the standardization of a field device interface as a basic requirement" also describes the most urgent concern of the industry, namely, to minimize the wiring effort caused by the rapidly developing automation technology and the resulting sharp increase in the number of field devices by means of a serial field bus [19]. When the 27 regional Profibus user organizations merged to form *Profibus & Profinet International* (PI) in 1995, there were already numerous systems and even more manufacturers of field devices with a Profibus interface. The field bus is now anchored in the IEC 61508 and IEC 61784 [12] standards, and there is probably no manufacturer of control technology or field devices that does not offer a Profibus interface for its products, at least as an option.[37]

Although the Profibus is still used in numerous applications, it is a bit outdated. Even if it is still completely sufficient in many applications thanks to its robust mode of operation and the now 12 Mbit/s data rate, it cannot even come close to keeping up with modern *Ethernet-based*

---

[37]Further information in [19] and many other publications.

bus systems. There are many reasons for this, but we would like to highlight just two.

First, it is the physical interface based on the standardized EIA-485 physics.[38] Its characteristics with a maximum of 32 participants, line structure with a maximum extension of 1,200 m, and a data rate of typically 10 Mbit/s[39] are no longer sufficient for today's requirements. Today's objectives are more oriented toward standard Ethernet with as unlimited resources as possible in terms of expansion, data rate, and network topology.

Another weak point of Profibus is the transmission according to the master-slave method without cross-traffic. In plain language, this means that the exchange of information is only possible between the master (e.g. the control CPU) and exactly one slave (the field device or a decentralized peripheral module with digital/analog I/O modules). The slaves are not able to communicate with each other, everything must first go to the master, and only the master can transfer this information to each individual slave participant. A motion application in which many slave axes working in a network require the position of the master axis at the same time and preferably without jitter is not possible in such a constellation.

This works better with the *CAN bus* (Controller Area Network). Originally, CAN was promoted by the automotive industry, which wanted to put a stop to the ever-increasing size of their vehicles' wiring harnesses in order to save weight as well as costs. This worked so well that not only the aviation industry but also the automation industry discovered this protocol for themselves, and with the founding of the *CAN in Automation* (CiA), higher layers were also added, such as the *CANopen* standard as European standard EN 50325 — 4 in 1995. The CAN bus is also based on EIA-485 physics and therefore has the same electrical properties as the *Profibus*, but the implied multi-master protocol does not require a central instance, and every participant can communicate with everyone. In addition, *CAN* works according to the producer-consumer concept (one sends, and all can listen in). With appropriate configuration and prioritization, axis networks can therefore be operated with a cycle time of up to 1 ms,

---

[38]EIA-485 only describes the physical interface and no data protocol. The same physics is used by e.g. also the CAN bus.

[39]But not with maximum expansion! With 32 subscribers and long cable lengths, you quickly end up with less than 100 kbit/s.

and the service-oriented CANopen standard makes work easier for programmers.[40]

Although the CAN bus with CANopen is still used successfully in many applications, its resources are no longer sufficient due to the increasing volume of data and the maximum of 31 possible participants in increasingly decentralized structures. The Ethernet,[41] which was established in the office world almost in parallel and at enormous speed from the 1990s onwards, attracted the automation world right from the beginning with its initially 10–20 times higher data rate. However, both the Ethernet physics and the most commonly used *TCP/IP protocol* have a number of disadvantages compared to the EIA-485 physics and the bus protocols which are based on it. We would like to take a closer look at just four:

**Network topology**
Ethernet does not support a direct *line topology*, as it is often required in automation. The individual devices are always connected to a *HUB* or *Switch* via point-to-point cabling. In a line topology, each participant therefore requires at least one 2-port HUB or switch with a third (internal) port.

**Cabling and plugs**
Ethernet cables are fitted with standardized IEEE802.3 plugs, which are designed for use in an office environment and are normally installed once and then remain in the device until they are disposed of. The ambient temperatures are "electronics-friendly" somewhere between 20 and 30°C, and unless coffee or cola is spilled, it is also dry and low-vibration. In an industrial environment, it is significantly warmer (control cabinet temperatures of over 50°C are not uncommon), vibrates, and is sometimes damp. Cables run through machines, lie on drag chains, and sometimes in a puddle of oil.

**Real-time capability**
The *CSMA mechanism* (Carrier Sense Multiple Acces) for resolving data collisions leads to irregular delays in data traffic in the Ethernet standard IEEE 802.3. And it gets worse. The switches used almost exclusively

---

[40]Further information on the CAN bus and CANopen can be found in [37] and many other publications.
[41]Ethernet largely corresponds to IEEE standard 802.3 and represents the physical layer (layer 1) and the data link layer (layer 2) in the OSI layer model.

today are designed to optimize data traffic in the network by intervening to control it. Therefore, they analyze the routes of the data packets that pass through their ports.

If there is suddenly an increased volume of data between two participants, these two are served with high priority, while the others have to wait for the end or a pause in this data stream.[42] That is due to the extremely disruptive *transmission jitter* in automation technology (Section 5.2.2) and is clearly a knock-out criterion for classic Ethernet from the office world.

**Data security**

There is a risk of unwanted, manipulated, or sabotaging data interfering with the device as soon as an Ethernet cable is plugged into a device and connected to a network. The danger is even greater if there is access to the Internet at any point in the network. What is not an issue at all in automation with *Profibus*, *CAN*, and other *non-Ethernet-based field buses* suddenly becomes common when production goods are integrated into a company network via Ethernet. Even the seemingly harmless use of the Ethernet connection of the CPU of a machine controller for remote diagnostics (which is very advantageous for error analysis during machine downtimes) is still rejected by production companies today for exactly this reason. Stuxnet[43] sends its regards!

These aspects alone have repeatedly and vehemently slowed down the triumphant advance of Ethernet in automation. It didn't help that this technology had proven itself millions of times over around the world and that the costs of Ethernet hardware were lower than those of field bus counterparts of automation.

Toward the end of the 1990s, drive technology experienced a powerful upswing due to the increasing availability of the necessary electronic components, and mechanical designers increasingly replaced the central main drives surrounded by complex gearboxes with decentralized

---

[42] You can easily try this out yourself by sending several pings in quick succession from its computer to another subscriber, which is best called "Some switches further away". You will notice that quite a long transfer time is required for the first ping and that this reaches its lowest value after just a few attempts. The devices have memorized this track and prioritize it from now on.

[43] Stuxnet is a computer virus and is said to have destroyed centrifuges in 2010 in order to sabotage the Iranian nuclear program. Since then, those types of malwares have been seen as a component of cyber warfare [43].

servo technology. This also increased the technological requirements for ever more sophisticated dynamics, as the machine, which had previously been equipped with a large, single speed-controlled main drive, was now equipped with many small servo axes, which should also work synchronously and with as much precision as possible. In the beginning, the solution was to connect a servo amplifier that was "only" speed-controlled to a field bus for parameterization and diagnostics and to simply continue to manage the setpoint via an analog signal from the axis controller of the control system. This was not an elegant solution, as the digital-to-analog conversion on the one hand and the analog-to-digital conversion on the other meant that the dynamics as well as the accuracy suffered. It very quickly became clear that only intelligent drives are capable of achieving the highest precision and speeds guarantee. But you can't do much with them without a powerful digital interface.

In the meantime, some manufacturers used the *ARCNET* (Attached Resource Computer Network), which was first introduced in 1977.[44] Like Ethernet and CAN, it only represents the transmission layer and not the application layer in the OSI model. Although it was developed at the same time as Ethernet, ARCNET was initially able to establish itself on the market because of the better token-passing access method. In this method, a master sends a token (identifiable data packet) on its journey and passes it on in a fixed sequence. This is set in the design process using a node ID on the individual station modules. Once a station has received a token, it takes the information intended for it and adds its own. It then sends the token on to the next station, which thus receives data from its predecessors as well as information from the master. This enables a certain *cross-traffic* on the one hand,[45] while, on the other hand, no collision slows down the speed of transmission so that even with the highest network load, a higher speed is achieved at 2.5 Mbit/s than with Ethernet networks that are four times faster. Even though, according to the *ARCNET Trade Association* (ATA), more than 22 million ARCNET nodes are installed worldwide today, and the bus even made it into small control systems in the 1990s, it no longer plays a role in the automation of machines because development has simply moved on.

---

[44]Further information on ARCNET in [38].

[45]However, cross-traffic is only limited to one direction, as information can only reach the respective successor. The producer-consumer concept as used by CAN is much better.

The psychological strain became so great around the turn of the millennium that the manufacturers began to seriously search for a powerful field bus. In the meantime, Ethernet had of course also developed further, and there was and still is no technical basis that could represent serious competition. Almost at the same time, numerous manufacturers, institutes, and organizations were attacking Ethernet in order to make it acceptable for the automation world. When the first products were introduced two years later, a veritable battle for the title of "My bus is the fastest!" began. The astonished users could hardly distinguish between true and false claims and to counter this, almost all manufacturers transferred their solutions to user organizations along the lines of the Profibus User Organization (PNO) or the CiA. But the fight did not stop, only the weapons changed. Now every single concept was put through its paces by countless institutes, companies, engineering offices, and universities, analyzed and compared with the others. In the end, the user was a lot smarter but still as clueless as before.

In the meantime, the smoke has disappeared, and the individual concepts have proven their advantages and disadvantages in different industries and numerous applications. Today, the right system for the right application can be selected with a high degree of certainty, and many publications such as the system comparison of the POWERLINK Standardization Group (EPSG [3]) can be used as a serious and reliable basis for analysis due to the easy verifiability of the statements.

And now the next revolution is upon us, but more on that later. At this point, let's summarize the most important properties of a field bus so that it meets current requirements:

**Sustainability of the investment**
- Ensuring upward and downward compatibility
- Simple infrastructure suitable for industrial use (cables, plugs, structural components, ambient conditions, etc.)
- Free choice of transmission media (cable, light, radio, etc.)
- Free choice of network topology (line, star, tree, etc.)
- Hotplug capability (automatic detection and registration of subscribers)
- Access to technology (user organization, licenses, etc.)
- Support through international standards

**Performance**
- Lowest possible cycle time
- Strictly deterministic time behavior with as little jitter as possible

- Synchronous and asynchronous data traffic with large data volumes
- Direct cross-communication, e.g. according to the producer-consumer principle (PubSub)
- Possibility of safety communication

**Economic efficiency**
- Costs of the individual components (individual nodes, infrastructure, etc.)
- Implementation effort (development, application, commissioning, etc.)
- Service and diagnostics (device technology, training, etc.)
- Operating costs (license costs)

### 5.4.2 Ethernet-based field buses: Properties and mode of operation

In order to make Ethernet fit for automation technology (see the requirements mentioned), two fundamental problems had to be solved — first, an infrastructure suitable for industry and second, ensuring real-time capability and deterministic.

The former is fairly easy to accomplish — at least that was the general perception. But the reality was quite different. Initially, it was thought that the components that had proven themselves millions of times over in the office world would also do their job sufficiently well in industry.

Users still fall for this fallacy today when they simply use the much cheaper Ethernet cables from wholesalers instead of the more expensive industrial cables and connectors. It is important to know that even the torsional moment of a simply twisted cable and the resulting tilting of the nearby connector are enough to cause contact failures after a short time due to a slight but permanent vibration, which are also extremely difficult to find.

However, deterministic is a much bigger problem. The Ethernet standard IEEE 802.3 uses the *CSMA method* to resolve data collisions, which by definition prevents reliable time behavior.

The explanation for that is simple: If two subscribers communicate with each other on the entire line and in the entire network, all other subscribers must wait until the line becomes free. If this is the case, everyone who wants to send something tries to gain access rights. This inevitably leads to collisions if several participants want to access the free line at the exact same moment. In this case, everyone retreats and waits for a randomly generated time until they try again. They repeat this process until one of them succeeds. The probability of multiple attempts for this procedure increases with each additional participant in the network so that the actual data rate in a network always depends on the participants and their network activities. The switches already mentioned are used in the office world to alleviate this problem. Therefore, a local segment with seven participants develops in an office with five PCs and two printers, for example. If something is to be printed from a PC, the switch will isolate the data traffic between the two devices involved so that the higher-level network is not affected at all. A switch only allows access to this network if a PC wants to access a remote sever, for example.

As already mentioned in the beginning, collisions of this kind are counterproductive for the automation of production systems, just like the use of simple switches. Avoiding collisions is the only way to use the widespread and cheap Ethernet chips and guarantee synchronous data traffic at the same time. For this purpose, three basic approaches have been established (Figure 5.43).

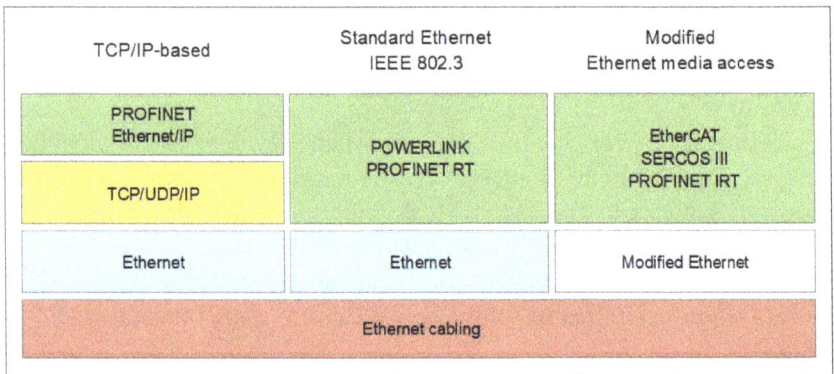

| TCP/IP-based | Standard Ethernet IEEE 802.3 | Modified Ethernet media access |
|---|---|---|
| PROFINET Ethernet/IP | POWERLINK PROFINET RT | EtherCAT SERCOS III PROFINET IRT |
| TCP/UDP/IP | | |
| Ethernet | Ethernet | Modified Ethernet |
| Ethernet cabling | | |

**Figure 5.43.**   Three paths lead to the real-time capability of Ethernet.

*Source:* According to [3].

## Based on TCP/IP

The protocols are based on the normal layers of TCP/IP and real-time mechanisms embedded in the top communication layer.

## Standard Ethernet

The protocols are based on the normal Ethernet layers. Such solutions benefit from the further development of Ethernet without additional investment.

## Modified Ethernet

The Ethernet layers, the mechanism, and the infrastructure of Ethernet have been modified. Such solutions prioritize performance over conformity.

The organization of data transmission and the way in which real-time behavior are established differentiates the industrial Ethernet systems. Two methods have established themselves for this purpose.

## Single frame procedure

With this method, individual telegrams are sent to the subscribers, who then also respond with individual telegrams. All other protocols mentioned are based on this procedure (Figure 5.44).

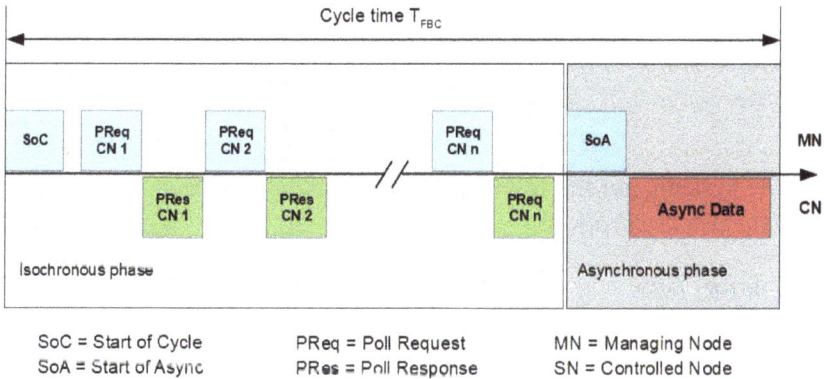

**Figure 5.44.**   Single frame procedure using the example of the POWERLINK.

*Source*: According to [3].

## Summation frame method

In each cycle, the data for all network nodes is sent in a telegram that passes through the nodes arranged in a ring topology one after the other and collects the response data at the same time. The summation frame method is used by Profinet IRT, SERCOS III, and EtherCAT (Figure 5.45).

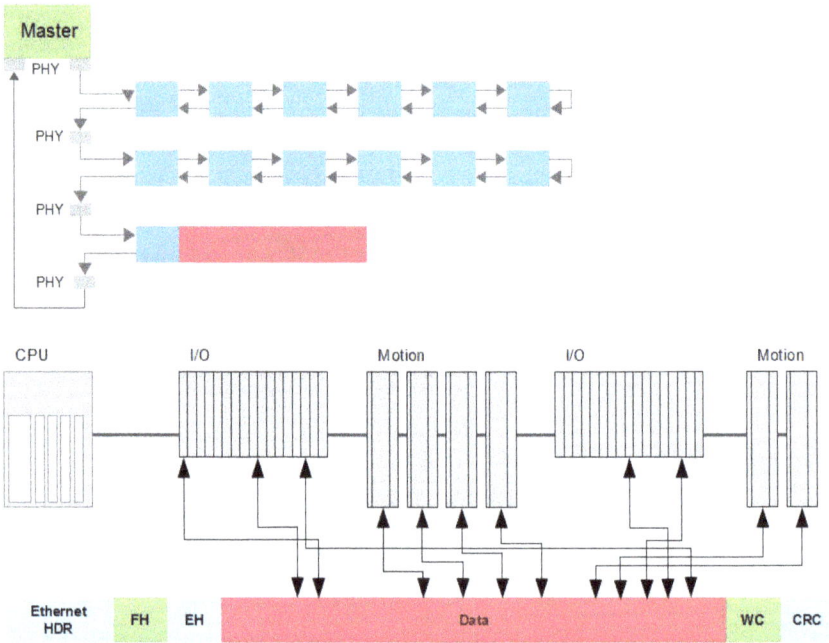

**Figure 5.45.**  Summation frame method using the example of EtherCAT.

*Source*: According to [3].

The systems use the following different methods for *network access* and *data synchronization*:

- *POWERLINK*
  The master grants the participants permission to send.
- *EtherCAT and SERCOS III*
  The master sets the clock for sending the summation frame telegrams before.
- *PROFINET IRT*
  Synchronized switches control the communication.
- *EtherNet/IP*
  CIP Sync is used to distribute time information in the network in accordance with the IEEE 1588 standard.

And which field bus is best suited for function- and object-oriented modularization?

We established in Chapter 2 that, with flexibility in mind, it must be possible to simply remove or add individual modules depending on the technological requirements without any further consequences for the rest of the machine or system. Consequently, the hot-plug capability, which refers to the automatic detection and registration of participants, is a crucial feature and a clear knock-out criterion for a modular concept in terms of Automation 4.0. Furthermore, the system should not be completely decommissioned during a converting, as this would extend the converting time unnecessarily. The use of modified interfaces is also not compatible with a flexible and open structure, just as the technologically required real-time capability and the transmission of safety-relevant data must be guaranteed in all cases.

Even though all the aspects mentioned at the beginning must be taken into consideration, it is the latter points that must be considered when selecting a suitable field bus for a modular system. Taking into account the analysis and evaluation in [3], we would like to take a closer look at the aspects mentioned here.

**Hotplug capability**

In principle, most suitable in this case are all systems which are not tied to a specific topology. The only exception is the *single frame method*.

The *summation frame method* already leads to limitations in the basic approach due to its mandatory ring topology. This does not change, even if various measures for a certain hotplug capability are implemented in the structures and protocols.

*EtherCAT* offers a certain hot plug capability because the slave controllers automatically close open ports if no connection is detected. But then the clocks distributed in the system are also not synchronized, which can affect some applications. However, this measure only works if there is exactly one gap. The removal of a second module at a different location leads to the creation of a segment which cannot be reached from either side.

**Modified interfaces**

It is irrelevant on which technical basis the field bus operates if machines should only be structured with integrable modules, provided that all other arguments are in favor of this field bus. The same principle applies to the bus used internally in a structure with autonomous modules. Externally, it would be unacceptable, as this would fundamentally question the sustainability of the investment of a modular structure.

**Real-time capability**

According to IEEE 802.3, the minimum length of an Ethernet frame is 512 bits, with a maximum of 12,208 bits. This means that at a data rate of 100 Mbit/s, between 6.06 μs (resulting from 5.1 μs + 0.96 μs for the inter-frame gap) and 122 μs are required for transmission. This data alone makes it plausible to understand the power of a field bus.

Let us assume, for example, that in an application with decentralized I/O nodes and electrically synchronized servo drives, frame lengths of 256 or 1024 bits are required per node and three devices are present in each case. A total *summation frame* would then have to have a net length of $3 \cdot 256 + 3 \cdot 1024 = 3840$ bits.[46] In addition, there is an overhead of approx. 50 bits and a frame would therefore be about 4 Kbit long, which means a throughput time of approx. 40 μs at a data rate of 100 Mbit/s. In addition, there are various delay times in the respective connections and the throughput time increases further if an area is reserved in the frame for the transmission of asynchronous or safety-relevant data. Consequently, it is clear that a $T_{FBC} \approx 100$ μs with the lowest Jitter[47] is actually possible with the summation frame method and a real application.

The *single frame procedure* would require at least twice the time for the same application, as each master telegram requires an adequate one from the slave. The data flow is also delayed by the infrastructure components. With a HUB, for example, this is typically 440 ns, which in the previously selected application means an additional delay of at least $13 \cdot 440$ ns = 5.72 μs with just one HUB. On the other hand, with a line topology, there is a separate HUB before or in each node so that, in the worst case, each HUB must be passed through twice for a single data exchange. In the example selected, this means a total of 12 HUB passages for the furthest station so that a total of 48 HUB passages generate an additional delay of approx. 21 μs for a complete cycle including SoC. Nevertheless, with this method, a $T_{FBC}$ of 200 μs is realistic, even if asynchronous data traffic is added.[48]

---

[46] The synchronous data volume for EtherCAT is a maximum of 1024 bytes [39].

[47] According to [40], this is around ±20 ns for EtherCAT.

[48] With Powerlink, the number of master telegrams can be reduced to one using the so-called multiplex procedure, thus achieving a considerable performance gain. Broadcasting also allows direct cross-communication, so that this disadvantage can be mitigated somewhat with the appropriate configuration.

**Safety-relevant data**

All the systems mentioned in [3] support safety communication in different ways. It is therefore necessary to consider how safety-relevant telegrams can be passed through to the individual safety components when selecting a field bus, especially for communication with autonomous modules. With *openSAFETY*, there is currently only one safety protocol which is suitable and freely available for all bus systems.[49]

It is certainly not possible to make a comprehensive recommendation for a specific field bus system at this point because, as always, all pros and cons depend on the specific application. Our recommendation is to specify the requirements and to make a well-founded decision based on the criteria already mentioned.

However, if you choose solutions that are based on standards[50] or are themselves open standards and still meet tough real-time requirements at the same time, you are very likely to be well equipped for future requirements.

Nevertheless, as we see in the following section, it may soon no longer be necessary to make a fundamental decision.

### 5.4.3  OPC UA in Industrial Ethernet

Increasingly large real-time networks in production plants pose major challenges for machine and plant operators, as the implementation of the *Industrial Internet of Things* (IIoT) leads to networks becoming more common which have several hundred nodes at field level. However, as the number of nodes increases, the engineering of the networks also becomes more complex, time-consuming, and expensive. In addition, different and incompatible protocols still have to be connected via bridges or gateways to form an overall network. If different machines from numerous manufacturers should communicate with each other in a typical machine hall for digital production, which in turn has different control systems with different field buses or industrial Ethernet networks, this heterogeneity leads to a great deal of effort during commissioning and maintenance. Even if these machines work separately from each other, numerous gateways and interfaces have to be programmed and maintained. This requires money, time and specialized personnel. At the latest when

---

[49] Further information on openSAFETY in [41].

[50] Since 2017, Powerlink has been the only internationally standardized Ethernet field bus (IEEE 61158) and openSAFETY is the only open and standardized safety protocol to date.

reaction times become necessary in the hard real-time range, e.g. to synchronize a robot and a plastic injection moulding machine for automated removal, this can no longer be implemented in a meaningful way.[51] Additionally, jitter-free communication between individual subsystems and components almost seems impossible due to proprietary solutions.

This is where the open standard *OPC UA* (OPC Unified Structure)[52] comes into play. OPC UA is a manufacturer-independent industrial M2M communication protocol which is characterized in particular by its ability not only to transport machine data (control variables, measured values, parameters, etc.) but also to describe it semantically in a machine-readable manner. The first version of the protocol was adopted in 2006 and has been published as the IEC 62541 series of standards since 2010. The multilayered service-oriented OPC UA structure provides basic services as abstract, protocol-independent method descriptions which are the basis for the entire OPC UA functionality and have currently been implemented by all major controller manufacturers [20]. The standard guarantees that machines with controllers from different manufacturers can be easily coordinated in one system. The protocol itself is also platform-independent and the communication stack can be ported to any operating system and embedded hardware. Therefore, OPC UA is the ideal I4.0-compliant communication protocol from the control level to ERP systems. All production process data, whether within a machine, between machines, or between a machine and a database in the cloud, can be transferred with OPC UA via a single protocol, making the classic field buses at factory level unnecessary.

Specifically, based on the RAMI 4.0 architecture model, [50] describes OPC UA for the individual communication levels of digital production as follows:

- The "product" network describes the interface of the manufactured asset itself, which is always product-specific. However, if the product is a machine, an OPC UA server interface is recommended.
- OPC UA Field eXchange — UAFX is recommended for the "Field Device" network.
- OPC UA with the client/server architecture is recommended for the "Control Device", "Stations", and "Work Center" network.
- OPC UA Pub/Sub via MQTT is recommended for the "Enterprise" network.

---

[51] See the examples in Chapter 1.

[52] Further information on OPC UA in [20] and other publications.

The recommendations make it clear that OPC UA is much more than a protocol: It is a collection of technologies. OPC UA offers a modern architecture for permeable information networks instead of a classic communication pyramid with many different protocols [50].

For this to work reliably, OPC UA transmits the data on the basis of information models, whereas classic bus systems transport dimensionless data, i.e. simple numbers without units or other information that must first be interpreted on the target system for further processing. We speak of a semantic interpretation of the data in this context. This procedure is reasonably practicable as long as machines work independently of each other. However, as soon as the data is to be transferred to other units — for example, other machines, SCADA, or ERP systems in a cloud — this semantic knowledge is lost, and the data are only values without dimension. In case of communication with another system in this way, the semantic descriptions are usually passed on in long tables and often entered into these systems by hand. The effort is immense, and the probability of errors is high. With OPC UA, this procedure is no longer necessary, because the OPC UA information models enable every participant to understand the data without any further explanation.

Let's take a sensor that measures a temperature of 5°C as an example. Conventional protocols transmit the value "5" to the controller as an integer data type. There, the transmitted number is stored as a temperature value in °C, which also has certain limit values. With OPC UA, the value "5" as well as the descriptive data is made available. In this case with the information that it is a temperature value that was measured in °C and in which limit values the value should be within. Other participants in the OPC UA network can now query and understand this information.

For example, if a new report should be generated in the ERP system, it can look up information in the network. This information can then be collected in a database and displayed in a report. Previously, this was only possible if a data transfer was programmed manually and the semantic information for each individual value was stored in the ERP system. If a variable was changed in the machine, it also had to be reprogrammed in the ERP system. The susceptibility to errors is reduced and flexible machine and system concepts can be implemented much more easily by using the information models of OPC UA [21, 22].

There are even more advantages to the OPC UA: The protocol provides implemented security functions with which data can be transmitted reliably and securely in order to connect the production and IT worlds. Moreover, the methods in OPC UA enable machine or system parts to call up certain functions in other machine or system parts. This not only makes it easy to implement networked digital production but also modularization in the sense of Automation 4.0.

To this end, [50] emphasizes the fundamental importance of OPC UA for Companion Specifications. While OPC UA defines how information can be exchanged securely, OPC UA Companion Specifications define what is exchanged as information. Information models based on the OPC UA metamodel (grammar and syntax) are defined and documented here. That is key to efficient digital data exchange. Combining standardized data exchange technology (OPC UA) with standardized semantics of the information which should be exchanged (Companion Specifications) leads to the following added values:

- Interface development costs are reduced for the machine or component manufacturer
- The integration of machines, components, and services at the machine operator is made easier
- The machine operator can process the information with higher availability and increased transparency without error-prone interpretation efforts.

OPC UA Companion Specifications also offer other advantages, such as

- **Domain-specific information models:** The semantic descriptions of their data and functionality are defined for assets, e.g. robotics, pumps, motors, auto-ID devices, industrial kitchen appliances, and laboratory analyzers, or
- **Protocol translations:** For existing industrial protocols, e.g. IO-Link and ProfiNET, it is defined how these are to be unambiguously converted to OPC UA, or
- **Engineering translations:** Implementation and mapping of PLC programming languages, such as IEC 61131-3.

The world's leading standardization organization in the plastics industry *EUROMAP*, for example, has defined the protocol as the basis of two EUROMAP interfaces. The umbrella organization of the packaging industry OMAC has also integrated it into the global standard *PackML*.

Everything that sounds that good, usually has a catch. This is also the case with OPC UA: Although the protocol is fast and suitable to use on small controllers and field devices due to its resource-saving coding, it is neither deterministic nor cyclical in terms of automation. OPC UA reaches its limits in processes with real-time requirements. It is therefore also unsuitable for intermodular communication if, for example, movements need to be synchronized. Therefore, the *OPC Foundation* and the *Industrial Internet Consortium* (IIC) [23] are working on two extensions that are intended to make OPC UA a real-time-capable communication standard. First, the protocol has been expanded to include a *publish/subscribe model* (Pub/Sub) and second, it will also be based on the *Time-Sensitive Networking* standard (TSN, IEEE 802.1).

We want to take a closer look at both and at the same time evaluate their benefits in a modular automation structure.

**Publish/Subscribe model**

OPC UA currently works with a *client/server mechanism*. A client requests information (request) and receives a response from a server (response). Among other things, this system reaches its limits when the network has many participants. A publish/subscribe system, on the other hand, enables so-called *publishers* (data sources and servers) to distribute selected data or services to any number of *subscribers* (data sinks and clients). To do this, the subscribers must have indicated their interest in specific information by registering.

The Publish/Subscribe model (Pub/Sub) thus allows *one-to-many* and *many-to-many* communication. A server publishes its data to the network and any client can participate and receive this data (Figure 5.46). This is similar to the so-called broadcast method based on the principle of "one sends and all listen".

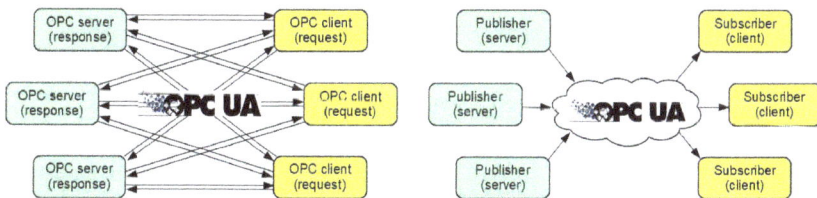

**Figure 5.46.** Principal mode of operation of client/server and publish/Subscribe model.

*Source*: B&R Industrial Automation GmbH.

This model is ideal for communication at line level as long as the machines in a factory work largely autonomously and only receive instructions or send status and diagnostic data. With Pub/Sub, it is also possible to implement operational automation technology, provided that the requirements for real-time capability are not too high and the size of the network remains manageable. But, machines, robots, and conveyor belts must communicate with each other in hard real time in order to achieve maximum productivity. OPC UA also reaches its limits with this method because processes have dynamic requirements in the single-digit millisecond range. It would be technically possible to make OPC UA itself real-time capable, but it would involve a great deal of effort and many disadvantages.

Numerous automation and robot manufacturers have therefore joined forces to take a different approach: OPC UA should be based on an extension of the Ethernet standard IEEE 802.1, *Time Sensitive Networking* (TSN).

**Time Sensitive Networking**
*Time Sensitive Networking* (TSN) refers to a series of sub-standards that have been standardized within IEEE 802.1 in order to make Ethernet fundamentally real-time capable. The automotive industry also relies on this standard, which is a major advantage of TSN. This means that the necessary semiconductor modules will be available very quickly and inexpensive in comparison.

And why Ethernet in the car? The amount of data transmitted in motor vehicles has increased exponentially in recent years, meaning that the bandwidth of previous bus systems is no longer sufficient. The full extent of this problem can be seen in mobile machinery. A typical system structure of large mobile machines involves up to 30 control units with several operating terminals and over 100 sensors and actuators communicating on several separate field buses for performance reasons. This can require a total of over 3,000 meters of cable weighing more than 100 kg. Reducing the number of cables alone would result in enormous advantages for production, reliability, service, and energy consumption.

Increasingly more media data also needs to be transmitted in cars. This can be video streams from a reversing or side camera as well as entertainment content that is played on the rear passengers' displays. As a first step, the automotive industry has therefore adopted the IEEE 802.1 AVB (Audio Video Bridging) standard, which enables synchronized and prioritized streaming of audio and video files. This allows data to be transmitted in a car via Ethernet.

The TSN initiative has developed from the AVB working group with the aim of reaching further industrial sectors and increasing the range of applications. The aim of the automotive industry is to also handle control tasks and functional safety applications via Ethernet. This requires cycle times in the real-time range and deterministic network behavior, and it is precisely these requirements that exist in the automation of modern production facilities. Initial implementations have been tested across manufacturers in testbeds under the direction of the IIC since mid-2017 and the results have been successively transferred to further developments. TSN is an extension of the Ethernet standard that has several improvements to make Ethernet real-time capable. It is based on fundamental standards that are described in [24–26]:

**Time synchronization according to 802.1AS-Rev**
This point is the basic prerequisite for the use of TSN in industrial real-time communication. The TSN sub-standard 802.1AS-Rev contains definitions of the *Precision Time Protocol* (PTP), which is responsible for the synchronization of all device clocks in the network.

**Precisely timed transmission of data packets and frames**
**(*traffic scheduling*)**
This includes the targeted control of data packets/frames in accordance with the TSN sub-standard IEEE 802.1 Qbv. A so-called Time-Aware Scheduler ensures that time-critical data always has priority and is not slowed down by general data traffic.

**Centralized, automated system configuration (*System Configuratlon*)**
This property is important for the automatic plug-in of a device. If a new device is added to the network, it automatically registers with a so-called Central Network Configurator (CNC). This establishes the communication relationship with the other devices with the corresponding requirements and reconfigures the network accordingly. The process is to be defined in the IEEE 802.1 Qcc standard in the future.

The functionality of these three methods has now been proven in various pilot products and testbeds. For example, a cyclical transmission time of 50 μs was achieved with PTP at a Jitter of less than 50 ns. This is even more precise than many field buses. This accuracy is sufficient for almost all industrial applications from line synchronization, the connection of ERP and HMI systems to highly dynamic control tasks with decentralized inputs and outputs and demanding motion applications.

The commissioning of components will also be massively simplified due to the new interoperability. All you have to do is plug in the network cable and communication will work throughout the entire network. Therefore, the combination of OPC UA TSN and Pub/Sub as a uniform standard has the potential to completely revolutionize communication not only above the control level, but together with TSN right down to the field bus level. According to [26], the age of Plug and Produce has arrived, and we have also come a lot closer to the USB principle in production [27] (Figure 5.47).

However, OPC UA TSN has the potential to compete with Ethernet-based field buses, especially in conjunction with the pub/sub method.

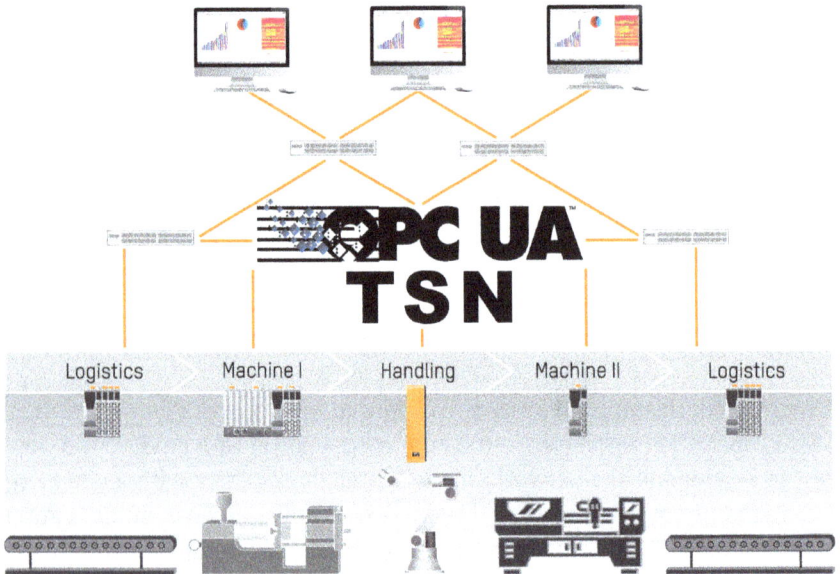

**Figure 5.47.** OPCUA TSN meets all the communication requirements of modern manufacturing.

*Source*: B&R Industrial Automation GmbH.

Even if the development of the standard has not yet been fully completed, it is already clear that, in the sense of Automation 4.0, *OPC UA TSN with Pub/Sub* is our favorite for modularized structures.[53]

Now we need to address the issue of secure transmission in terms of safety and security.

### 5.4.4  Single Pair Ethernet

*Single Pair Ethernet* (SPE) is an extension of Ethernet with the aim of also reaching sensors and actuators via Ethernet. The development is still relatively new but is already seen as the next generation of communication structure in automation. It was originally developed for automotive applications and now promises nothing less than an end-to-end connection from the sensor to the cloud. And that in almost every application, whether that is in industry, logistics, buildings, or wherever data is generated [51].

SPE basically follows the approach that the high data transmission rates of the IT world are not required for coupling with various field devices but longer cable lengths with a compact design and simple and robust cabling are. It does contradict the constant "faster and shorter" of the IT world by having the lower data transfer rates. However, many applications do not make full use of the capacities already available, especially intelligent field devices. These are currently still connected via proprietary field bus, IO-Link, ASi, and gateways, sometimes over several levels. With Industry 4.0 and the smart factory, however, these system breaks can no longer be maintained. They require end-to-end networking of the three levels of the automation pyramid. This is possible with SPE, as it enables a continuous Ethernet connection and also offers considerable technical and economic advantages [51, 52]. These include the following:

- Networking with TCP/IP without system interruptions
  - ○ Each field device can be addressed via IP
  - ○ Any Ethernet technologies and protocols such as OPC UA, TSN, or openSafety are possible
- Replacement for most proprietary field buses up to IO-Link[54]

---

[53] Almost all relevant manufacturers are already working on implementing this technology in their systems, meaning that proprietary field buses will gradually become less important in the foreseeable future.

[54] Detailed information on this can be found at [42].

- Long distances up to 1,000 m
- Power supply for end devices via the same line
- Economical due to savings in material (only one wire pair), installation space, flexibility, and easier and error-free installation thanks to pre-fabricated cables and connectors.

The standardization of Single Pair Ethernet is in full swing and manu-facturers of field devices and control components are also working on implementing this technology in their systems to replace proprietary systems.[55]

### 5.4.5 Secure communication: Safety

Communication in modular and flexible automation concepts not only involves *functional* but also *safe* data exchange — both at the control and line level of complex production systems. As a result, it is necessary to go beyond the individual (modular) machine and design the *safety technology* flexibly and enable safety line automation. Until now, it has been common practice for machines from different manufacturers to be con-nected directly in the machine hall to form a safety network with a great deal of programming effort. If, during operation, the machines, i.e. machines are removed or new ones are added, the safety technology must be reprogrammed and checked each time.

In addition, the new Machinery Directive will make *industrial secu-rity* mandatory from 2027, and in the future, only machines whose safety mechanisms cannot be manipulated even when integrated into a network will receive a CE mark.[56]

In regard to this aspect, OPC UA technology also offers a solution because it enables self-organizing I4.0-compliant safety networks in combination with a suitable safety protocol such as the open-source *openSAFETY*.[57] With this technology, it will be possible to remove or add modules or entire machines from the network without having to reprogram the safety technology. Self-validating machine lines are also

---

[55]Detailed information on this can be found at [52, 53].
[56]Detailed information on this can be found in [45, 46].
[57]Detailed information on this can be found in [47].

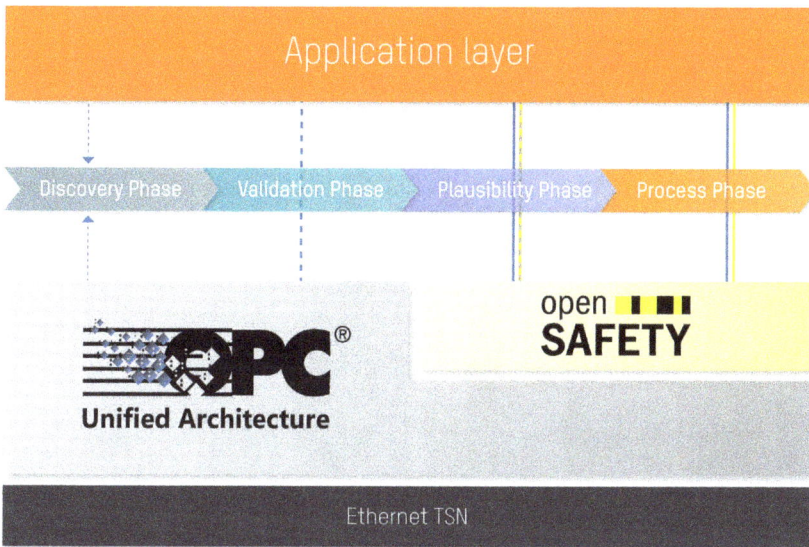

**Figure 5.48.**   openSAFETY used for data exchange the Publish/Subscribe mechanism of OPC.

*Source*: B&R Industrial Automation GmbH.

conceivable according to [28]. However, in order for the safety network to be able to organize itself and meet all security and safety requirements at the same time, a number of precautions are necessary, for which the embedded mechanisms in *OPC UA and openSAFETY* already form the basis (Figure 5.48).

The mode of operation can be illustrated in four steps (Figure 5.49), [28]:

**Discovery phase**
When a new device — i.e. a machine, a machine module or even a robot — is connected to a network, a secure connection is first established using the *OPC UA security mechanisms*. The new device then searches for other servers that offer safety functions. The OPC UA mechanisms *Discovery* and *Server Capability* are used for this. The OPC UA browsing services are then used to determine which functions and which attributes these servers offer. By doing this, each OPC UA server obtains a complete picture of the network without having to program a single line of code.

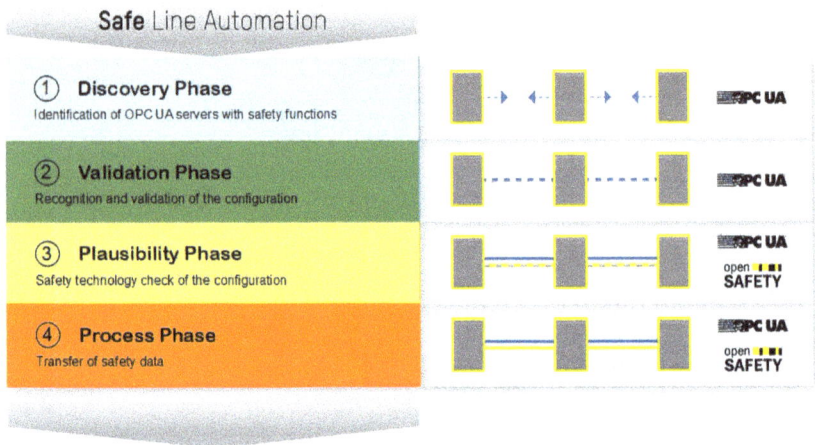

**Figure 5.49.** Phase model of a self-organizing safety network with OPC UA and openSAFETY.

*Source*: B&R Industrial Automation GmbH.

### Validation phase

The safety application now checks whether the new component is already known or whether all properties are equivalent to a previously validated configuration from a safety point of view. If this is the case, no further actions by the machine operator are necessary. If relevant differences are detected, the user is asked to confirm the correctness of the configuration using a standardized query via the visualization. The entries are saved permanently so that the new line configuration is automatically recognized in the future.

### Plausibility phase

In this step, each individual component checks whether the current configuration is plausible. This process is identical to the checks that are carried out each time a machine is switched on. It also tests whether the reaction and cycle times are sufficiently short to trigger the required safety reactions reliably.

### Process phase

Once all previous test steps have been completed, the exchange of safe process data starts via openSAFETY and the line network can start the intended production operation. If an emergency stop switch is now actuated in the network, all devices in the openSAFETY network are automatically informed. Each device then decides independently whether this information is relevant for its function and reacts accordingly.

The procedure described also supports the configuration of a *linear profile*, which even enables the object-oriented mode of operation described in Chapter 2. Machine or system parts communicate their status directly to their neighbors. If a machine part switches to a safe state, the direct neighbors decide independently whether they must also switch to a safe state or whether they can continue working — possibly at a reduced speed. In the end, the entire line communicates with each other without the intervention of a higher-level system or a human being.

The exchange of safe process data, known as *Safety Process Data Objects*, using the OPC UA protocol and the Pub/Sub-mechanism also guarantees that openSAFETY nodes can communicate directly with each other and therefore achieve very short reaction times. However, data exchange during plausibility checks requires data queries in the form of *Safety Service Data Objects*. These use OPC UA method calls to avoid unnecessary data load in the networks and on the OPC UA servers. In principle, the openSAFETY safety protocol can also use any field bus and any industrial Ethernet network as a transport medium.

In addition, the *black channel principle*[58] enables the exchange of safety-related data without the transport protocol being able to influence the secure data [28].

The example shows that safe communication in terms of safety and the defined guiding principles of Automation 40 is possible with the technologies already available today.

### 5.4.6 Secure communication right into the cloud: Security

The topics of *Big Data* and *increasing the efficiency* of (digital) production were presented as important topics for the future in Section 1.2.9. Comprehensive and meaningful data of the entire production process is required in order to meet these requirements, and to avoid the feared flood of data, the data must be aggregated in the individual systems. This means

---

[58] The black channel principle allows the transmission of secure and non-secure process data via the same networks, regardless of the regular data transport mechanism used on this line.

the detailed pre-processing of technological and process-relevant data. Therefore, numerous manufacturers offer so-called *edge devices* which are either integrated into control systems or incorporated into local field buses as a supplement. They are the last physical instance before the cloud or the network and logically use all OPC UA instruments.

*Edge controllers* are therefore able to directly access and pre-process the data generated in the control system, such as energy and vibration measurement values and technology-based production data, as required. This reduces the load on the control system in two ways. On the one hand, it can be equipped with less computing power, and it also requires less memory capacity. This means it can be smaller and more cost-effective. Whereas production data was previously only stored in controllers for a short time and then deleted to make room for more up-to-date data, the edge controller now takes over this task. It not only aggregates the data but also stores it until it has transferred everything to a higher-level cloud. This reduces bandwidth requirements and costs for the cloud, and no data is lost if the connection is interrupted.

In [29], we differentiate between three edge variations, which illustrate the basic working methods and how they are offered by different manufacturers (Figure 5.50).

**Edge connect**
This method is used if individual measured values should only be transmitted at longer intervals. If, for example, a sensor picks up a signal every hour, it may be useful to send the data directly to the cloud. For such simple applications, an Edge Connect solution, in which unprocessed I/O signals are encrypted and sent to the cloud via OPC UA, is sufficient.

**Edge embedded**
This refers to the integration of a purely software-based edge controller, which makes sense when sufficient resources are available in a control system and/or the amount of data to be processed remains manageable. In this case, the controller itself is embedded into the network or with an additional module via Ethernet and OPC UA.

**Edge controller**
These devices are relevant for larger data volumes and are integrated into a control system as stand-alone devices with corresponding firmware (e.g. on an industrial PC) or as an add-on module.

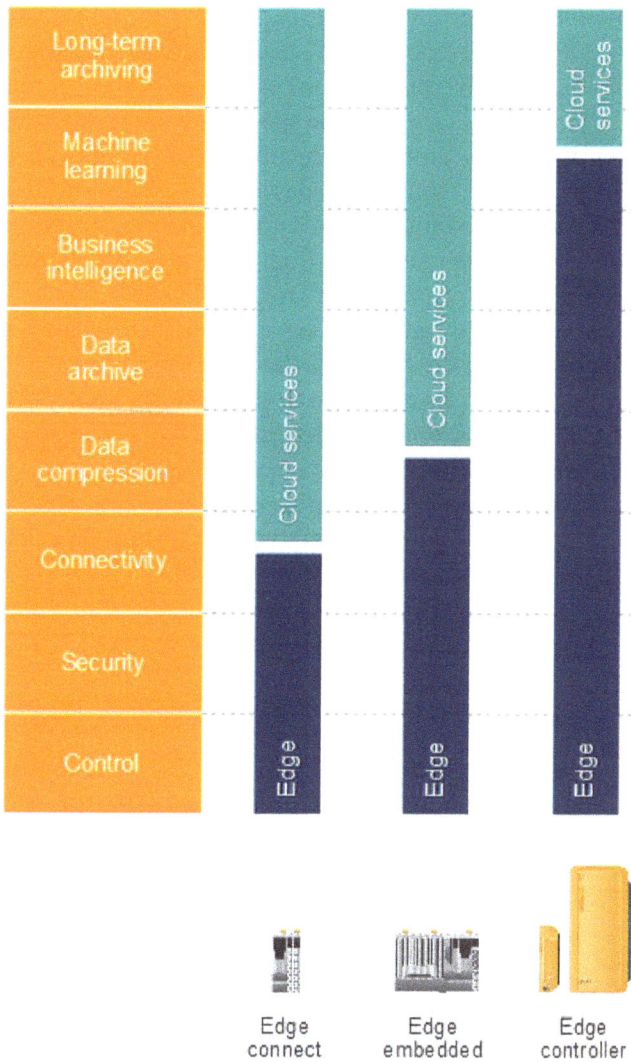

**Figure 5.50.** Three basic edge architectures offer solutions for every application.

*Source*: B&R Industrial Automation GmbH.

Conveniently and in the spirit of Automation 4.0, the transfer to a cloud should take place via OPC UA. This has the advantage that a cross-manufacturer connection is understood both in the IT world and of almost all hardware and software at the control level.

And this has another advantage: The security mechanisms already built into the OPC UA protocol can be used for secure data transmission. These are, for example, the functions for user identification or role assignments with which access rights can be specifically assigned or restricted.

However, to reliably protect production systems from cyber attacks, further IT measures are required, as described in detail in [30]. Only a staggered system of firewalls as well as perimeter and buffer zones can reliably protect production networks from sabotage and manipulation by malware such as viruses, worms, and Trojans (Figure 5.51).

Additional hardware gateways ensure that individual systems or machines can be specifically opened for access for remote maintenance or diagnostics, for example, as described in [30] (Figure 5.52).

A service technician, for example, uses a *Link Manager* (VPN-like access) to establish a connection to the machine and to gain access to the machine or system from a PC or smartphone. The central point is a *Gate Manager* (central administration portal), which contains the machine pool management in which the user accounts, authorizations, and machines are

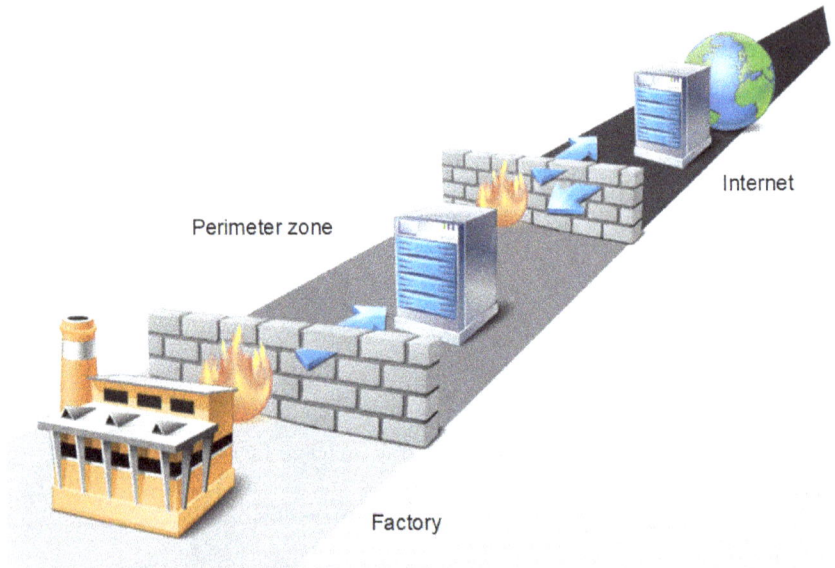

**Figure 5.51.**    Maximum security through double firewalls: Data from a production plant is transferred to the perimeter zone and is only available there for access from outside.

*Source*: B&R Industrial Automation GmbH.

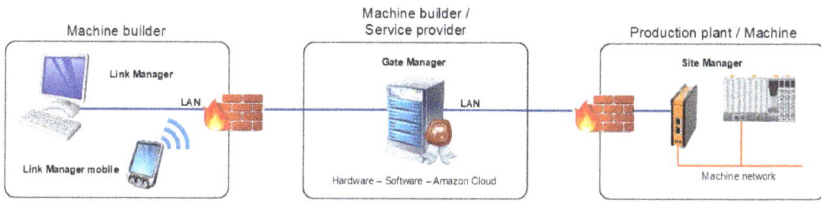

**Figure 5.52.** Site Managers act as hardware gateways for secure access to industrial systems.

*Source*: B&R Industrial Automation GmbH.

managed. It establishes the connection between technicians and machines after checking the access authorizations.

The *Gate Manager* is a virtual machine and is set up to meet the specific requirements of a system. It can be operated by a cloud provider or set up on the machine manufacturer's own server.

Finally, a *Site Manager* (additional hardware) establishes the connection between the machine or production network and the Internet. A firewall is integrated into the Site Manager to ensure maximum security and communication with the Internet exclusively takes place via encrypted web protocols. The *Patch Management* of the firmware of all Site Managers is also extremely important in order to close any potential security gaps immediately.

These measures provide the best possible protection of the production facilities while simultaneously providing authorized access as best as possible.

## 5.4.7 Summary

The explanations in this chapter are titled "Communication is (almost) everything". But in the age of the Internet of Things, digital production, and the individualization of production facilities through modular systems, is communication truly only about almost everything?

The answer to this question is a clear NO and this should be made clear in this chapter. The focus is increasingly on communication today and in the future, apart from the fact that modular automation can no longer function without it. Future technological developments will determine both the possibilities and ultimately also the requirements for modern and, above all, suitable industrial communication. The examples of

OPC UA TSN or the standardization activities surrounding Industry 4.0 show that the developments are in full swing or already fixed in standards. Therefore, current and future automation solutions must be measured by their ability to support these developments or drive them forward with intelligent approaches. Manufacturers who believe they can secure their market position with proprietary individual solutions will have to rethink their approach or else will fall behind.[59]

Consequently, designers are well advised to rely exclusively on standardized and as far as possible, open communication protocols and solutions in current and future developments right from the start. Controllers that are upwards and downwards compatible, HMI systems that communicate web-based via OPC UA, and system-internal field buses that (if required at all) are at least capable of transporting these data packets in hard real time and together with safety-relevant information are certainly the right approach in terms of Automation 4.0.

## 5.5  Adaptive and intuitive: HMI 4.0

The design of a *human–machine interface* (HMI) or *user interface* (UI), has some special features in contrast to the other components of an automation system. Those features result from its special positioning: the direct contact with people. These are the parts of a production system with which they come into intensive contact, from where they receive their information and via which they issue commands. And this is where the problem begins.

People have the ability to compare their own behavior patterns with their environment — an interlocutor, a politician, or a technical system — and to expect the same. This can be seen most clearly in the development of service robots. Numerous studies have shown that these are most readily accepted when they not only resemble humans in their external appearance but also in their behavior. It is therefore not surprising that the more logical and intuitive the design of an HMI, the better and, above all, the more error-free the communication with a production system. This includes both the external appearance and the way in which information is provided and

---

[59]This sentence was written in the 1st edition in 2018. Just five years later, it is already proving to be true, as the first companies have already lost their established market position for precisely this reason.

operating actions can be carried out. This means that communication with a production system must be tailored to the sensations and expectations of the operator in the best possible way. For example, studies have shown that humans can only perceive a maximum of seven pieces of information at the same time [48]. In practice, however, operator terminals are literally over-loaded with information. This automatically overwhelms the operator, which leads to rejection, stress and ultimately misinterpretation and incorrect operation.

On the other hand, the operator immediately trusts the HMI system if he encounters one that resembles the familiar interface of his smartphone and offers him information in a clearly structured form. However, both component suppliers and system developers have to counter this challenge resulting in the fact that HMI systems are subject to even greater innovation dynamics than the other components of an automation system.

Additionally, each module in a modular system, whether integrated or autonomous, requires more or fewer HMI components for the respective functions. However, this means that the design and content of the HMI can change with each module configuration. If, for example, a gluing station is added to our perfect binder, it must also be possible to visualize and operate it directly on the HMI of the backbone.[60] If this module also has a local HMI, the user interfaces must harmonize with each other. This means that it must be irrelevant for the local controller from which device module-related operating actions are performed. This also includes, for example, a service technician's smartphone or tablet that is registered and authorized in the company network.

We want to take a more general look at the world of HMI and evaluate established solutions in terms of their suitability for modular production systems before we present a solution approach in terms of Automation 4.0.

### 5.5.1 Meaning and basic tasks

The operability of a production system determines its economic success to at least the same extent as the design because what use do the most sophisticated technological functions have if they are not operable or are difficult to operate?

---

[60]This is usually the main operating point of a production system.

The cockpit of a small car alone has around 100 different control elements: key switch, steering wheel, pedals, radio, and even displays for speed, mileage, and much more. And there are certainly many people who have not even seen all the displays before selling their car. For this reason, designers and usability specialists are trying to develop ever more sophisticated and intuitive operating concepts so that the customer feels like they can manage everything from the first glance. The same applies to a production system.

In Chapter 1, we discussed the fact that production systems are becoming increasingly complex, technologically more extensive, and more flexible. Accordingly, the amount of information and intervention options required for effective operational management of the production system is also increasing. The automated changeover to different product formats alone requires extensive parameterization. Moreover, there are various operating modes and numerous settings that can be changed by the operator during operation which is based on information made available to him from various displays and signals. Therefore, one task of engineering is to provide the operating personnel with the *relevant information* at the *right time* so that they can *intuitively* carry out the appropriate operating actions.

*Relevance* means that it is not important to provide operators with as much information as possible because unfortunately, much not really help a lot. On the contrary: When there is too much information at a glance, the truly important things can go unrecognized, leading to incorrect decisions.

*Timely correct* means that important information is provided at the right time or in the right context. For example, there are critical states where action must be taken immediately or as quickly as possible. This can occur, for example, if a quality parameter drifts away or a raw material container is almost empty. However, if the machine is in setup mode, the empty container message would be unsuitable. Nevertheless, it is necessary if the production should start.

The information must be clear, meaningful, and logical in order to make the *right intuitive decision*. The message "Machine faulty" is a typical message which does not lead to intuitive action. However, if it reads "Standstill due to defective main fuse" and this message is supplemented with a picture and instructions for action, then it can be assumed that the fault will be rectified quickly.

However, caution is also required. When automation engineers think about the human–machine interface, they primarily think about text displays, operator terminals, and industrial PCs. However, it is often forgotten that the HMI includes every single adjusting screw, every hand-wheel, signal lights, switches, and everything else that is important for their usage in production or service. Therefore, it also requires economic common sense to combine all options with discretion instead of cramming everything into a multi-touch UI down to the last light or adjusting screw.

For this reason, clarifying *which information* the operating personnel need and *when* and *what actions* are required is the first step in the design process of a user interface. The corresponding technical are determined in the second step. On the one hand, this includes the hardware, which includes all device technology right down to the hand crank or adjusting screw. Only then are the HMI components and functions of the automation technology defined.

The developers receive most of the specifications for this process from the requirements specification, whereby the content of the UI is subject to constant further development during the development. The specific content, from the individual image content of a display to the text of the last error message, will only be determined at the end of development, although the essential content and design specifications can be integrated into the requirements specification.

In summary, the tasks of human–machine communication can be described as follows:

- General user guidance
- Display of process statuses and alarms (display, indicators, LED, etc.)
- Input operator commands (touch, buttons, switches, rotary encoder, etc.)
- Input/output of data (USB, storage media, scanner, RFID, etc.)
- User administration (login/out, etc.)
- Intelligent support for operational management (condition monitoring, AI, etc.)

## 5.5.2 Constructive design

According to [49], the characteristics of HMI components (automation technology) can be simplified as follows:

- **Simple displays and operating elements**
  For example, individual signal lights or traffic lights, switches, buttons, and sliders
- **Text and numeric displays and keyboards**
  For example, large displays, alphanumeric displays, and keypads

- **Display panels in terminal mode**
  For example, control panels with or without integrated keypad
- **Embedded controller (combination of HMI and PLC on one hardware)**
  For example, intelligent control panels with or without integrated keypad
- **PC-based systems**
  For example, panel PCs (IPC combined with (touch) display and possibly additional keypads) or IPC in the control cabinet with remote display on the machine
- **Intelligent communication systems (acoustic, visual, and tactile)**
  For example, virtual reality, voice-based input and output, image recognition, identification via RFID or barcode scanner, touch-sensitive surfaces, and much more.

As already mentioned, only one main or many distributed control points may be required in a production system for optimum operation depending on the technological task. How these can be implemented and combined with each other is described in detail in [49] and is not discussed any further. Rather important for the HMI of a modular production system is the software-side implementation.

### 5.5.3  SCADA system

According to the definition, a *SCADA system* (Supervisory Control and Data Acquisition) is an HMI in combination with an IPC. However, the boundary between an industrial PC, an embedded PC, and a control CPU is very fluid so that the defined assignment of HMI plus IPC can no longer be clearly drawn. *SCADA software* can therefore be integrated more or less extensively on all platforms and with different technologies.
The following aspects must be considered in detail.

#### 5.5.3.1  *System classification*

There are many possible combinations of SCADA system software with the corresponding hardware and operating system. The following perspectives help with classification:

- **The SCADA application works on a control CPU**
  In this constellation, their processing must be subordinate to the real-time system and the technological control functions must not be

impaired. This is usually ensured by the SCADA components using the idle time of the cyclical program sequence. However, this also means that it shares resources with a cyclical network management and other non-real-time-critical tasks, for example.

- **The SCADA application works on an HMI panel with RTOS**
  As a rule, control programs are also processed on the panel in addition to the SCADA application. This means that the same application criteria apply as before, with the difference that an HMI panel usually has less computing power. Corresponding products are offered by numerous manufacturers.

- **The SCADA application works alone on an IPC or embedded controller**
  In this case, the hardware resources, in addition to the operating system, are available solely to the SCADA system. Generally, operating systems adapted for industrial use, such as Windows 10 IoT or similar are used. There are limitations with embedded controllers, as these usually have less CPU performance. Products from different manufacturers are used as SCADA software, which offer corresponding packages and drivers for communication with various automation systems for different industries.

### 5.5.3.2 *Engineering of SCADA applications*

It is also worth taking a look at the way in which a SCADA system is integrated. If, for example, control system manufacturers have integrated this functionality into their systems, the HMI application is usually configured using the same development tool that is required for engineering the PLC programs. This means that the developer usually has access to all data, usually with direct access to the variable pool and without restrictions.

However, if the SCADA system is either installed on an IPC alongside the runtime system of the PLC or provided via a separate HMI panel, this direct access is usually not available. In this case, the organization of the data required for the HMI sometimes causes considerable additional work and dynamic restrictions if, for example, real-time data for a trend is to be recorded in the HMI. Moreover, changes to data on one side which are not reflected on the other cause more errors to occur.[61]

---

[61] In modular systems, special attention must be paid to this issue.

The engineering of the HMI also includes the coupling between the PLC and SCADA system. If both platforms work on one system and this has a direct interface to a panel or terminal, the visualization (image output on the panel or terminal) and operation (evaluation of the pointing and button commands) take place directly via the drivers and resources of its own hardware (PC or embedded controller). However, the transfer of either the data from the control component to the HMI or the SCADA visualization to the HMI controller must be configured separately if both work on separate hardware. Part of the engineering is assessing the effort with which that is done.

Usually, editors are offered for the design of the visualizations, with which the selection and placement of the image elements can be carried out using the drag & drop method. This also includes defining their properties, including data assignment. The screen elements themselves are offered in numerous libraries which, depending on the industry, SCADA product, and license model, are either part of a standard package or must be purchased separately. Their scope can sometimes also be supplemented by self-created elements.

Web-based operating concepts are more interesting and much more future-proof. Figure 5.53 shows how they work in principle. In this concept, a web server works in the control CPU, to which freely edited web pages are added and whose important feature is that information can be called up and displayed with different tools (browsers) regardless of operating system variants. The required data can be called up as required from any terminal, PC, laptop, or web-enabled device on the Intranet from a web page on the machine. A web interface thus raises access to automation to a neutral, reliable, and universal level of information. Web technology also makes it possible to offer HMI components of automation systems from different manufacturers on any end device.

In terms of system technology, interactive websites can also be used to operate a machine or system. If the pages are dynamized with JavaScript elements and the data connection is made via OPC UA, an HMI is created that can run on all current and web-enabled end devices such as smartphones, tablets, or panels (Figure 5.54). Among others, this concept offers the following advantages:

- **The operating and service personnel only need one terminal device**
  For example, an entire production line with machines from different manufacturers can be operated and diagnosed using WiFi and a tablet.

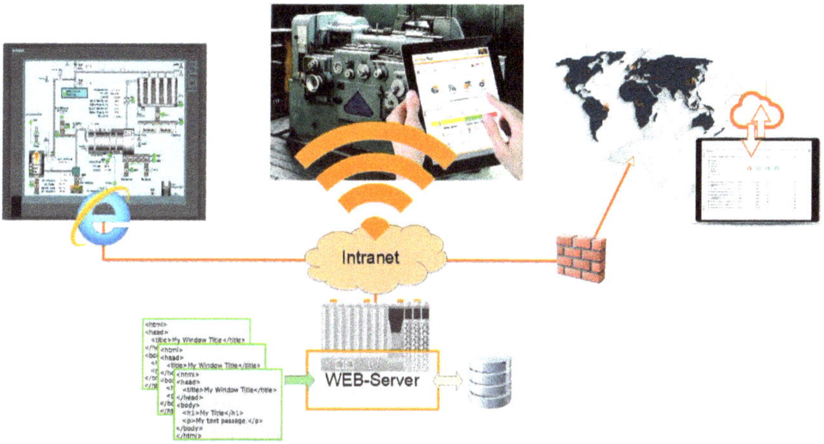

**Figure 5.53.** PLC-integrated web servers serve local, mobile, and external HMI devices with visualized data.

*Sources*: ©B&R 2020 and ©Siemens 2020. All rights reserved.

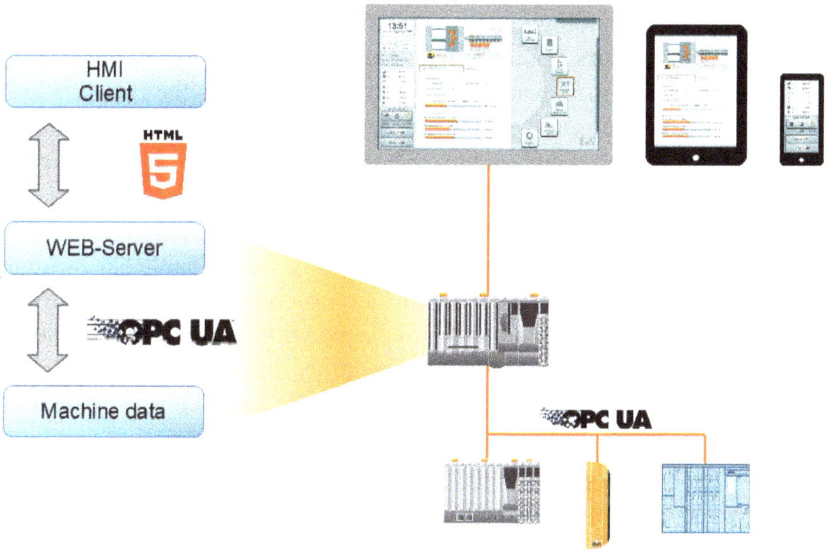

**Figure 5.54.** A web-based HMI concept makes it easy to use any end device.

*Sources*: ©B&R 2018 and ©Siemens 2018. All rights reserved.

This is particularly advantageous in widely branched systems, as there is no need for a permanently installed HMI device at every necessary operating position.

- **Unlimited service and diagnostics worldwide**
  Web technology makes remote diagnostics easier than ever. Provided data security measures are in place, service can be carried out via the Internet from anywhere in the world. The service technician can receive an alarm message from home and carry out operating actions himself. The manufacturer's service personnel can also use this technology to assist with troubleshooting, for example, without having to be on site.

- **Modular components provide their own HMI functions**
  The HMI can be combined on a common terminal in a modular production system with different mechatronic units from different manufacturers and other automation technology. The operator does not even have to notice this if, for example, he presses a button to change pages and is then on the HMI of a machine module. Web-based HMI and OPC UA-supported data procurement also significantly simplify the integration of very complex supplier components with their own automation technology into an overall system.

- **Arbitrary editors create more flexibility**
  It is completely irrelevant how and with which tools the HTML pages are created. They can be created with a simple text editor or with any licensed or open-source editor. The only requirement is compliance with the conventions applicable to the respective target system.

- **Open and future-proof**
  On the one hand, the technology allows any web-capable hardware to be used on any side. Ideally, there are no compatibility problems, if either the controller or a panel is replaced with a newer model at some point during the life cycle. On the other hand, *HTML*, *OPC UA*, *VBScript*, and *JavaScript* are standards that are likely to remain in place for a very long time. This means that subsequent further development can build on existing versions, even if other automation technology is to be used.

The explanations make it clear that the web-based design of a user interface is not only future proof and sustainable but also appears to be the most suitable for the modular design of an automated production system.

# References

[1] DIN EN ISO 13849 — 1:2016-06, Safety of machinery — Safety-related parts of control systems — Part 1: General principles for design (ISO 13849 — 1:2015); German version EN ISO 13849 — 1:2015.

[2] Siemens AG: SINAMICS S120 — System overview, 2018.

[3] EPSG — Ethernet POWERLINK Standardiziation Group: Industrial Ethernet Facts, 3rd edn., Fredersdorf 2017, available online at: http://www. ethernet-powerlink.org/fileadmin/user_upload/Dokumente/Industrial_ Ethernet_Facts/EPSG_IEF3rdEdition_e.pdf, last accessed: 17.01.2018.

[4] Sick AG: Detection of paper edges, Waldkirch 2018, available online at: https://www.sick.com/de/de/branchen/druck/press/papierfuehrung/ detektion-der-papierkanten/c/p350010, last accessed: 23.01.2018.

[5] B&R Industrie-Elektronik GmbH: User manual for control systems 2003/2005/2010 (B&R industrial electronics product family B&R2000), Eggelsberg 2004.

[6] B&R Industrial Automation GmbH: reACTION Technology, 2018, available online at: https://www.br-automation.com/de/technologie/reaction-technology/, last accessed: 24.01.2018.

[7] Schmertosch, T.: Drastic reduction of reaction times down to 1ms in modular mechanical engineering and decentralized automation structures, in: Tagungsband VVD, Dresden 2015.

[8] Schmertosch, T.: Funktionale Sicherheitstechnik für Maschinen und Anlagen, in: Beitrag des VEMAS Industriearbeitskreis AUTOMATION im Forum der INTEC, Leipzig 2015.

[9] Official Journal of the European Union L 157/35, Directive 2006/42/EC of the European Parliament and of the Council of May 17, 2006 on machinery and amending Directive 95/16/EC (recast), Brussels 2006.

[10] DIN EN ISO 13849 — 2:2013-02, Safety of machinery — Safety-related parts of control systems — Part 2: Validation (ISO 13849 — 2:2012); German version EN ISO 13849 — 2:2012.

[11] DIN EN 61508 — 1:2011-02; VDE 0803 — 1:2011-02, Functional safety of electrical/electronic/programmable electronic safety-related systems — Part 1: General requirements (IEC 61508 — 1:2010); German version EN 61508 — 1:2010.

[12] DIN EN 61784 — 3:2017-09; VDE 0803 — 500:2017-09, Industrial communication networks — Profiles — Part 3: Functionally safe transmission for Field buses — General rules and specifications for profiles (IEC 61784 — 3:2016); German version EN 61784 — 3:2016.

[13]  Official Journal of the European Union C 87/1, Communication from the Commission in the framework of the implementation of Directive 2006/42/EC of the European Parliament and of the Council of 17 May 2006 on machinery, and amending Directive 95/16/EC (recast), Brussels 2012.

[14]  Federal Ministry of Labor and Social Affairs (ed.): Interpretationspapier zum Thema "Wesentliche Veränderung von Maschinen", published by the BMAS on 09.04.2015 — IIIb5-39607-3 — in GMBl 2015, No. 10, p. 183 — 186, Berlin, 2015.

[15]  World of Print: Saddle stitcher Stitchmaster ST 450 achieves increased productivity with new feeders and new stream feeder, Ratingen 2010, available online at: http://www.worldofprint.de/2010/03/08/sammelhefter-stitchmaster-st-450-erreicht-mit-neuen-anlegern-und-neuem-streamfeeder-gesteigerte-produktivitaet/, last accessed: 24.01.2018.

[16]  Pilz GmbH & Co. KG: Safety Compendium (Version 11.16, 01.12.2017), available online at: https://www.pilz.com/download/open/TechBo_Pilz_safety_compendium_1004669-DE-01.pdf, last accessed: 31.01.2018.

[17]  B&R Industrial Automation GmbH: Integrated Safety Technology. User manual, Eggelsberg 2017.

[18]  B&R Industrial Automation GmbH: Integrated safety technology (overview brochure), 2018.

[19]  Profibus Nutzerorganisation e. V. (PNO): Homepage, https://de.profibus.com/community/die-profibus-nutzerorganisation/, last accessed: 06.02.2018.

[20]  OPC Foundation: Homepage, https://opcfoundation.org/, last accessed: 06.02.2018.

[21]  B&R Industrie-Elektronik GmbH: Practical test passed: OPC UA becomes real-time capable (B&R industrial electronics customer magazine), in: *Automotion*, No. 11/2016.

[22]  B&R Industrial Automation GmbH: OPC UA TSN — From the field level to the cloud (B&R Industrial Automation customer magazine), in: *Automotion*, No. 09/2017.

[23]  Industrial Internet Consortium (IIC): Homepage, https://www.iiconsortium.org/, last accessed: 06.02.2018.

[24]  Industrial Internet Consortium (IIC): OPC UA TSN — A new Solution for Industrial Communication, Whitepaper 2018, available online at: http://www.computer-automation.de/uploads/media_uploads/documents/1516349639-226-opc-ua-tsn-a-new-solution-for-industrial-communication.pdf, last accessed: 21.02. 2018.

[25] B&R Industrial Automation GmbH: TSN-the turbo for OPC UA?, in: *Automotion*, No. 11/2015.

[26] B&R Industrial Automation GmbH: The age of plug-and-produce has arrived, in: *Automotion*, No. 11/2017.

[27] Thielicke, R.: Industrie 4.0: Fraunhofer-Institut arbeitet am USB-Prinzip für die Fertigung, in: Technology Review v. 4. Dezember 2013, online available at: http://www.heise.de/-2059877, last accessed: 24.06.2017.

[28] B&R Industrial Automation GmbH: Automating lines safely, in: *Automotion*, No. 11/2016.

[29] B&R Industrial Automation: From the edge to the cloud, in: *Automotion*, No. 09/2017.

[30] B&R Industrial Automation GmbH: Protecting process control systems from cyber attacks, in: *Automotion*, No. 09/2017.

[31] World of Print: Saddle stitcher Stitchmaster ST 450 achieves increased productivity with new feeders and new stream feeder, Ratingen 2010, available online at: http://www.worldofprint.de/2010/03/08, last accessed: 24.01.2018.

[32] B&R Industrial Automation GmbH: Stitchmaster ST 500, https://www.youtube.com/watch?v=TZ1eTWzxllI, last accessed: 24.01.2018.

[33] Pilz GmbH & Co. KG: Online Lexicon, 2018, available online at: https://www.pilz.com/de-DE/knowhow/lexicon, last accessed: 31.01.2018.

[34] Sick AG: Product portfolio, 2018, available online at: https://www.sick.com/de/de/c/PRODUCT_ROOT, last accessed: 31.01.2018.

[35] Pilz GmbH & Co. KG: Product overview (Version 5.01, 01.11.2017), available online at: https://www.pilz.com/download/open/Leaf_Product_overview_1004011-DE-02.pdf, last accessed: 31.01.2018.

[36] Pilz GmbH & Co. KG: Control technology brochure (Version 10.59, 01.04.2017), available online at: https://www.pilz.com/download/open/Leaf_Product_overview_1004011-DE-02.pdf, last accessed: 31.01.2018.

[37] CAN in Automation (CiA): Homepage, https://www.can-cia.org/, last accessed: 08.02.2018.

[38] ARCNET Trade Association (ATA): Homepage, www.arcnet.de/, last accessed: 08.02.2018.

[39] EtherCAT Technology Group: EtherCAT Overview (Version 5.2), available online at: https://www.ethercat.org/download/documents/ETG_Brochure_DE.pdf, last accessed: 21.02.2018.

[40] Beckhoff Automation GmbH: EtherCAT — The real-time Ethernet Field bus, 2017.

[41] EPSG — Ethernet POWERLINK Standardiziation Group: Homepage, https://www.open-safety.org, last accessed: 08.02.2018.

[42] IO-Link consortium: Homepage, http://www.io-link.com/de/, last accessed: 08.02.2018.

[43] Holland, M.: Stuxnet angeblich Teil eines größeren Angriffes auf kritische Infrastruktur des Iran, available online at: https://www.heise.de/security/meldung/Stuxnet-angeblich-Teil-eines-groesseren-Angriffs-auf-kritische-Infrastruktur-des-Iran-3104957.html, last accessed: 25.07.2018.

[44] Berufsgenossenschaft Rohstoffe und chemische Industrie-Technische Sicherheit (BG RCI, ed.): Informationsschrift "Wesentliche Veränderung von Maschinen", Stand 01.09.2020, Langenhagen, 2020, available online at: https://www.bgrci.de/fileadmin/BGRCI/Downloads/DL_Praevention/Fachwissen/Maschinensicherheit/Informationspapier_Wesentliche_Ver%C3%A4nderung_von_Maschinen.pdf, last accessed: 25.07.2023.

[45] Regulation (EU) 2023/1230 of the European Parliament and of the Council of 14 June 2023 on machinery, and repealing Directive 2006/42/EC of the European Parliament and of the Council and Council Directive 73/361/EEC, available online at: https://eur-lex.europa.eu/legal-content/EN/TXT/?uri=CELEX:32023R1230, last accessed: 02.08.2023.

[46] Pilz GmbH & Co. KG: Whitepaper on Industrial Security, Version 2018-01, available online at: https://www.pilz.com/mam/pilz/content/uploads/wp_security_de_2018_10.pdf, last accessed: 02.08.2023.

[47] openSAFETY: Homepage, online: https://www.br-automation.com/de-de/technologie/opensafety/, last accessed: 05.08.2023.

[48] 1st VEMAS Innovation Network Mechanical Engineering Saxony. Integration of people into production. Operability of machines and systems. 2nd fireside chat of the AUTOMATION industry working group. Lichtenwalde: s.n. 2015.

[49] Schmertosch, Th.: Structured automation systems: The right component selection for modular machines and systems. Vogel Communications Group, Würzburg 2021.

[50] VDMA e.V. (April 11, 2023). Discussion paper — Interoperability with the asset administration shell, OPC UA and AutomationML: Target image and recommendations for action for industrial interoperability. Retrieved July 10, 2023 from https://www.vdma.org/viewer/-/v2article/render/78243357.

[51] Single Pair Ethernet — The network infrastructure for the Industrial IoT. Whitepaper, Weidmüller Interface GmbH & Co KG (ed.), November 2019, online: https://www.weidmueller.de/de/loesungen/single_pair_ethernet/index.jsp, last accessed: 05.09.2023.

[52] Single Pair Ethernet. Whitepaper, U.I. Lapp GmbH (ed.), November 2019, online: https://www.lapp.com/de/de/produkte/eigenschaften-technologien/ technologien/single-pair-ethernet/e/000211, last accessed: 05.09.2023.

[53] Single Pair Ethernet System Alliance, Homepage: https://singlepair ethernet.com, last accessed: 05.09.2023.

# Chapter 6

# Automation 4.0 at a Glance

The term Automation 4.0 illustrates how deeply the requirements of Industry 4.0 intervene in the development and manufacturing of production systems and the production processes themselves, and how comprehensive modularization can generate market advantages for manufacturers and operators. It also shows that with a consistent object-oriented approach and working method, the production process can be designed efficiently and error-free for manufacturers and operators.

We have shown how modular production systems can be described as objects in the form of technical assets using several examples. Moreover, we have applied methods of object-oriented software development and structured project planning to the development and design of mechatronic systems and supplemented those with practical aspects of automation technology.

The observations show that autonomous modules can be integrated flexibly into a machine and system concept because they do not require central control for operational control and regulation tasks, are completely encapsulated in terms of function and design, communicate in a service-oriented manner only with the partners defined in the ERD and do not require any knowledge of the overall system. We have also shown that production systems can only work in an environment that is compliant with 14.0 if they communicate via standardized interfaces.

Further points of discussion were the Ethernet and the OPC UA bus protocol in its real-time-capable version OPC UA TSN. In addition to technological service-oriented communication, this communication

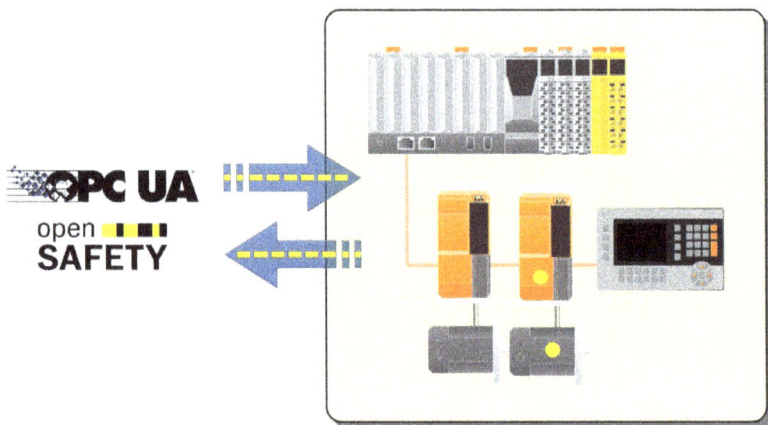

**Figure 6.1.**    I4.0-compliant automation of a production system.

protocol can also be used for safety-related communication with the open-Safety protocol.

Figure 6.1 shows an example of a model of a production system configured in this way (to which the definition of an autonomous module also applies) with control, drive, and safety technology as well as an operating panel as an HMI component.

The installation of a module-internal administration shell is also required if the integration of a module or a production system with the CP Class 44 is planned in a digital production (Figure 6.2).

If production systems are modularized according to the described method in a functional and object-oriented manner, functional modules are created, allowing for the flexible customization of a production system with minimal effort. This method also enables the operator to individually configure the system. It can easily be expanded and modified if the requirements change. The machine modules themselves go through a completely decoupled life cycle so that the manufacturer's innovations can be introduced to the market without affecting the rest of the product portfolio.

Finally, it was shown which boundary conditions need to be considered when applying this concept in practice. These include the requirements for upward and downward compatibility of the control hardware and software, the necessary responsiveness during operation and the execution of safety-relevant functions as well as the implementation of

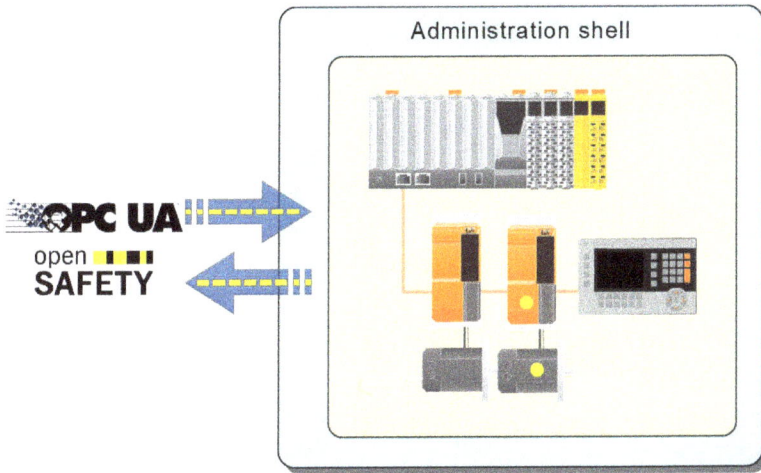

**Figure 6.2.**    I4.0-compliant automation of a production system (CP Class 44).

internal and cross-modular requirements for operation and process visualization. Finally, we concluded the topic with several considerations regarding the protection of data traffic (IT security), digital project planning, and integrated and AI-supported quality assurance.

The authors hope that this will turn *Automation 4.0* into a successful concept.